Planet Earth
ATMOSPHERE

Other Publications:

THE CIVIL WAR
COLLECTOR'S LIBRARY OF THE CIVIL WAR
LIBRARY OF HEALTH
CLASSICS OF THE OLD WEST
THE EPIC OF FLIGHT
THE GOOD COOK
THE SEAFARERS
THE ENCYCLOPEDIA OF COLLECTIBLES
THE GREAT CITIES
WORLD WAR II
HOME REPAIR AND IMPROVEMENT
THE WORLD'S WILD PLACES
THE TIME-LIFE LIBRARY OF BOATING
HUMAN BEHAVIOR
THE ART OF SEWING
THE OLD WEST
THE EMERGENCE OF MAN
THE AMERICAN WILDERNESS
THE TIME-LIFE ENCYCLOPEDIA OF GARDENING
LIFE LIBRARY OF PHOTOGRAPHY
THIS FABULOUS CENTURY
FOODS OF THE WORLD
TIME-LIFE LIBRARY OF AMERICA
TIME-LIFE LIBRARY OF ART
GREAT AGES OF MAN
LIFE SCIENCE LIBRARY
THE LIFE HISTORY OF THE UNITED STATES
TIME READING PROGRAM
LIFE NATURE LIBRARY
LIFE WORLD LIBRARY

FAMILY LIBRARY:
HOW THINGS WORK IN YOUR HOME
THE TIME-LIFE BOOK OF THE FAMILY CAR
THE TIME-LIFE FAMILY LEGAL GUIDE
THE TIME-LIFE BOOK OF FAMILY FINANCE

This volume is one of a series that examines the
workings of the planet earth, from the geological
wonders of its continents to the marvels of its
atmosphere and its ocean depths.

Cover
A brilliant rainbow graces a stormy sky over
the Rocky Mountains in Alberta, Canada.
Rainbows can be seen only at a precise angle
relative to the sun's rays: The arc's beauty
is literally created in the eye of the beholder.

Planet Earth

ATMOSPHERE

By Oliver E. Allen
and The Editors of Time-Life Books

Time-Life Books, Alexandria, Virginia

Time-Life Books Inc.
is a wholly owned subsidiary of

TIME INCORPORATED

FOUNDER: Henry R. Luce 1898-1967

Editor-in-Chief: Henry Anatole Grunwald
President: J. Richard Munro
Chairman of the Board: Ralph P. Davidson
Executive Vice President: Clifford J. Grum
Editorial Director: Ralph Graves
Group Vice President, Books: Joan D. Manley
Vice Chairman: Arthur Temple

TIME-LIFE BOOKS INC.

EDITOR: George Constable
Executive Editor: George Daniels
Director of Design: Louis Klein
Board of Editors: Dale M. Brown, Thomas A. Lewis,
Martin Mann, Robert G. Mason, John Paul Porter,
Gerry Schremp, Gerald Simons, Rosalind Stubenberg,
Kit van Tulleken
Director of Administration: David L. Harrison
Director of Research: Carolyn L. Sackett

PRESIDENT: Reginald K. Brack Jr.
Executive Vice Presidents: John Steven Maxwell,
David J. Walsh
Vice Presidents: George Artandi, Stephen L. Bair,
Peter G. Barnes, Nicholas Benton, John L. Canova,
Beatrice T. Dobie, James L. Mercer, Paul R. Stewart

PLANET EARTH

EDITOR: Thomas A. Lewis
Designers: Donald Komai, Albert Sherman
Chief Researcher: Pat S. Good

Editorial Staff for *Atmosphere*
Associate Editor: John Conrad Weiser (pictures)
Text Editors: William C. Banks, Sarah Brash, David
Thiemann
Staff Writers: Tim Appenzeller, Jan Leslie Cook,
Stephen G. Hyslop, Paul N. Mathless
Researchers: Jean Burke Crawford and Donna Roginski
(principals), Susan S. Blair, Sheila M. Green, Blaine
Reilly Marshall, Rita Thievon Mullin
Assistant Designer: Susan K. White
Copy Coordinators: Victoria Lee, Bobbie C. Paradise,
Diane Ullius
Picture Coordinator: Donna Quaresima
Editorial Assistant: Caroline A. Boubin

Special Contributor: Cathy Gregory (research)

Editorial Operations
Design: Arnold C. Holeywell (assistant director);
Anne B. Landry (art coordinator); James J. Cox
(quality control)
Research: Jane Edwin (assistant director),
Louise D. Forstall
Copy Room: Susan Galloway Goldberg (director),
Celia Beattie
Production: Feliciano Madrid (director),
Gordon E. Buck, Peter Inchauteguiz

Correspondents: Elisabeth Kraemer (Bonn); Margot
Hapgood, Dorothy Bacon (London); Miriam Hsia,
Lucy T. Voulgaris (New York); Maria Vincenza
Aloisi, Josephine du Brusle (Paris); Ann Natanson
(Rome). Valuable assistance was also provided by:
Wibo van de Linde (Amsterdam); Mirka Gondicas
(Athens); Helga Kohl (Bonn); John Howse
(Calgary); Lois Lorimer (Copenhagen); Jane Pender
(Fairbanks); Robert W. Bone (Honolulu); Judy
Aspinall, Millicent Trowbridge (London); Cheryl
Crooks (Los Angeles); John Dunn (Melbourne);
Christina Lieberman, Cornelius Verwaal (New York);
John Scott (Ottawa); Mimi Murphy, Ann Wise
(Rome); Katsuko Yamazaki (Tokyo).

For information about any Time-Life book, please write:
Reader Information
Time-Life Books
541 North Fairbanks Court
Chicago, Illinois 60611

Library of Congress Cataloguing in Publication Data
Allen, Oliver E.
    Atmosphere.
    (Planet earth; 8)
    Bibliography: p.
    Includes index.
    1. Atmosphere. I. Time-Life Books.   II. Title.
III. Series.
QC861.2.A35 1983   551.5   82-16768
ISBN 0-8094-4336-8
ISBN 0-8094-4337-6 (lib. bdg.)

THE AUTHOR
Oliver E. Allen, a former Time-Life Books editor, is the author of volumes in the Epic of Flight, Library of Health and Seafarers series. Research for a book on windjammers fostered his interest in winds and storms and provided background information for this volume.

THE CONSULTANTS
Richard A. Anthes, Director of the Atmosphere Analysis and Prediction Division at the National Center for Atmospheric Research, has also served as a research meteorologist at the National Hurricane Research Laboratory and as a professor in the Department of Meteorology at Pennsylvania State University. He has written more than 30 publications, including, with three colleagues, the authoritative textbook *The Atmosphere.*

Vilhelm Bjerknes is a meteorologist at the Climate Analysis Center of the National Weather Service. In his choice of a career, he thus follows in the distinguished footsteps of his father, Jacob, and his grandfather, Vilhelm, the founder of modern meteorology.

CONTENTS

THE PROTEAN SKY

The sky presents an eternally unfolding spectacle. One moment puffs of cumulus cloud skitter across it; the next, a billowing thunderhead perhaps 10 miles high may loom over the horizon. The 19th Century British author John Ruskin described the aerial drama as "almost human in its passions, almost spiritual in its tenderness, almost divine in its infinity."

Over the past four centuries, scientists have done much to explain the mechanics underlying this ceaseless pageant. Today, virtually every significant process and trend within the earth's shroud of air is monitored very closely. Weather satellites orbiting some 22,000 miles above the earth record the progression of air masses, the distribution of water vapor and the fluctuation of temperature at different altitudes. Laser beams and radar probe the clouds from below to locate areas where precipitation is forming. Balloons hoist sophisticated instrument packages as high as 18 to collect data on winds, the effects of man-made pollution and the behavior of the jet streams, which course through the atmosphere at speeds of up to 300 miles per hour.

Yet the prodigious wealth of information from these various sources has raised a host of new questions: If dust plays a critical role in the world's rainfall, what would be the effect of a stupendous volcanic eruption? What explains the observed connection between dark spots on the sun and intensified Arctic storms? What will be the long-term effects of the processes that deposit automobile emissions on the remote glaciers of Antarctica? Given these fresh lessons in the atmosphere's intricacy, even a scientist who knows that a rainbow is a straightforward consequence of meteorological optics can gaze at its gentle magic with a sense of wonder.

Benign despite its menacing appearance, an enormous altocumulus cloud looms in the sky over Argentina. Such clouds, called lenticular because of their lenslike shape, form in the turbulence on the leeward side of mountains.

Fog, which is simply an earthbound stratus
cloud, engulfs San Francisco and the Golden
Gate Bridge. Great banks of fog roll into
this area from the Pacific each summer as warm,
low-lying air chills rapidly over cool water.

8

The northern lights, properly called the aurora borealis, shimmer in swirling bands over an Alaskan forest. Such displays are actually gases that glow when particles emitted by the sun strike the upper atmosphere.

Signaling the end of a summer shower, a
rainbow arches over Kootenay Lake in Canada.
Rainbows form when sunlight strikes drops
of water falling from a rain cloud; the droplets
function as miniature prisms and split the
light into a curving spectrum.

Chapter 1

AIR'S ELUSIVE ELEMENTS

The British writer Roger Pilkington was whiling away a summer evening aboard a motor launch moored in the Baltic harbor of Svendborg, Denmark, idly watching a glow in the nighttime sky through an open hatch. His casual interest suddenly heightened, he wrote in 1962, "when the pattern began to move and flicker" in a way that could not possibly be accounted for by the lights of the waterfront. Scrambling erect for a better view, he confronted an awesome sight.

"I found the vault of the stars wonderfully lit with a general glow that became intensified towards a point about halfway up from the horizon," he recalled. "This area grew brighter and redder, almost as though aflame, bands of purple light darting out from it to vanish in a flickering green. Slowly, a little more to the westward and lower down, giant folds began to form and ripple like a gorgeously colored cloth of shot silk stirred by the wind. Rays like those of searchlights pierced the sky to the right, striking fanwise from the infinity of space. The whole sky was now alive with movement, and I watched it for more than an hour before the curtain began to fade imperceptibly. The rays dimmed, trembled and vanished; the deep rosy patch high above the wooded hills died, flared once again in lesser brilliance, and was swallowed at last in darkness like the dying embers of a glowing log."

Pilkington had witnessed one of nature's most magnificent spectacles— the northern lights, more properly known as the aurora borealis (the southern counterpart is the aurora australis). The aurora, named after the Roman goddess of the dawn, has mystified and bedazzled observers since the beginning of time. The Roman philosopher Seneca wrote two millennia ago of its "gaping displays, some of a flitting and light flame-color, some of a white light, others shining, some steadily and yellow without eruptions or rays." And in the Bible a passage in Maccabees describes the specter of "horsemen charging in midair, clad in garments interwoven with gold." The folklore of Eskimo peoples, who may see the strange lights as often as one night out of three, abounds in myths associating the glow with the departed: According to one tale, the northern lights are cast by lanterns carried by spirits guiding new arrivals to their world.

Scientists have long since dismissed such fearful analogies with precise explanations. But in a real sense the scientific account is fully as wondrous as the notion of some spiritual celestial activity. It speaks of a solar wind of tiny charged particles wafting through interplanetary space, enveloping the earth in an ethereal bombardment; it describes an immense magnetic

The lurid glow of a red veil—one of the eeriest of the auroras—shrouds the South Pole station of Showa in July of 1959. The time exposure that allowed the camera to capture the veil's bloody hue has distorted the images of the stars and the building's floodlight.

shield, generated by mysterious processes deep within the earth's core, that turns aside this invisible barrage, which, if undeflected, would soon end all life on the planet; and it explains how a few particles of the solar wind find gaps in the protective shield over the polar regions and collide with the molecules of the outermost reaches of the atmosphere to set off a series of reactions—akin to those that are harnessed in fluorescent lights—that generate the ghostly auroral luminescence.

Far more than a stupendous light show, the display Pilkington watched was one manifestation of the shimmering, multilayered collection of gases, liquids and particles that is the earth's atmosphere. From its origin as a searing mix of hydrogen, ammonia, methane and water vapor, the atmosphere has evolved into a complex global envelope of a multitude of constituents with a myriad of functions whose composite effect is to nurture life on the planet. It serves the complex chemistry of life as it distributes the vital gaseous elements that are the basis of plant and animal tissue; its onion-like structure, extending nearly 200 miles into space, protects the earth and its inhabitants from bombardment by missiles ranging from the tiny, invisible bits of solar wind to the occasional meteor flaming across the sky; its continuous internal shifting and seething—the winds—mediate the scorching heat of the equatorial latitudes and the numbing cold of the polar reaches.

The evolution of this marvelous matrix took four billion years; human knowledge of its dimensions and intricacies has come about principally in the past 400 years. And not until the entire gossamer structure of the atmosphere's dynamics had been sensed by a worldwide array of instruments linked by rapid communications, or, better still, had been seen from space, could the wonderful unity that underlies the processes be perceived. Without such a global perspective, primitive people and even early scientists were limited to musing about the seemingly fickle wanderings of nearby winds and clouds, or about the distant, obscure causes of a season of rain or a decade of drought. Merely to identify the atmosphere—to label its many constituents and grasp how they are arrayed around the globe—required sophisticated instruments and the vehicles to carry them aloft. And a full comprehension of the subtle system that accounts for the wind and the climate proved to be one of modern science's most taxing challenges.

The earliest knowledge of the atmosphere and its ways was garnered from long and hard experience by the farmers and the sailors who lived and died according to its vagaries. In the Eighth Century B.C., the Greek poet Hesiod advised his fellow farmers, "Take heed what time thou hearest the voice of the crane from the high clouds uttering her yearly cry, which bringeth the sign for plowing and showeth forth the season of rainy weather, and biteth the heart of him that hath no oxen." He also told sailors that it was safe to go to sea when the topmost leaves of the fig tree had attained the size of a crow's foot. Such perceptions still guide Greek seamen.

Natural philosophers of ancient Greece began keeping systematic records and used the patterns of the past to predict the future. Thales of Miletus achieved considerable prominence in 585 B.C. by successfully predicting a solar eclipse. He also comprehended the basic nature of what is now known as the hydrologic cycle, in which water evaporates into the air to form

A suffocating cloud of dust churns over a parched field in the Texas-Oklahoma panhandle during the 1930s. Droughts have recurred on America's Western plains every 20 to 22 years, a frequency that some climatologists see as linked to sunspot cycles.

clouds which produce rain that returns to earth. But, in a pattern that would often recur, when Thales turned to logical theory he went badly awry. He thought, for instance, that the earth was a disk floating in a universe of water.

Thales may have been the first to find a practical application for atmospheric knowledge. Teased by friends about the lack of rewards his wisdom had brought him, Thales is supposed to have resolved to defend his reputation. From his observations of weather trends, he concluded that the next olive crop around Miletus—on the eastern shore of the Aegean, in present-day Turkey—would be a good one. He bought all the olive presses in the vicinity, and when the bumper crop came in, Thales became rich overnight. Then, having made his point and being unimpressed by worldly wealth, he abandoned commerce and returned to the contemplative life of a philosopher.

No doubt influenced by Thales' brilliance, three other scholars who lived near Miletus, and who possessed curiously similar names, subsequently arrived at some important conclusions about the atmosphere. One of them, a student of Thales' named Anaximander, reasoned that wind was not some mysterious phenomenon regulated by the gods, but rather a natural "flowing of air." A fellow student, Anaximenes, announced his belief that air was the fundamental element of the universe and could become water or earth when compressed. Anaxagoras, who lived almost a century later but studied Thales' writings, hit on the truth that heat causes air to rise and understood that clouds result when rising air cools. His suggestion that temperature drops with increases in altitude, while it seemed self-evident to casual observers, was not entirely accepted in scientific circles until late in the 19th Century.

But after these promising beginnings, the study of the atmosphere was guided by some of the greatest thinkers of all time into a monumental dead end, where it was to stagnate for millennia. The mishap had its roots in the love of logic that transfused the thought of Classical Greece—to the benefit of philosophy but with disastrous results for practical wisdom. The Greek disdain for observable evidence and preference for symmetrical theory led to flights of fancy about the atmosphere that, despite their lack of scientific usefulness, occasionally contained fascinating glimmerings of truth.

Empedocles, a Sicilian poet-philosopher who cultivated a reputation as a mystic and prophet, proposed in the Fifth Century B.C. that air was one of four elements—the others were earth, fire and water—from which all things were derived. Sometimes, he said, these substances might combine to produce others, but generally they were opposed: Hot, dry fire ushered in summer, but when cold water gained the upper hand, winter arrived.

Nearing the end of his life, Empedocles resolved to disappear without a trace in order to foster the notion that he had been transported to heaven and made a god. He is alleged to have attempted to accomplish this by jumping into the crater of Mount Etna; sadly, the volcano spewed up his bronze sandals, marring the perfection of the act.

Empedocles' ideas—and his errors—were expanded during the Fourth Century B.C. by the renowned Greek philosopher Aristotle. Each of the four elements had its place in the natural order, said Aristotle, and motion took place when an element tried to return to its rightful position. The earth was in the center of things; water lay above it, air above that and fire above all. So an object composed of an earthen material, a rock, for example, would fall if raised. Rain, too, would fall, but fire would rise. The four elements, he decreed, were restricted to the earth's environs; the heavens were composed of "ether."

A few of Aristotle's concepts have proved to be surprisingly accurate. In a discourse on rain, for instance, he elaborated on the idea that air cools as it rises: "The efficient, controlling, and first cause is the circle of the sun's revolution. The earth is at rest, and the moisture about it is evaporated by the sun's rays and the other heat from above and rises upwards; but when the heat which caused it to rise leaves it, the vapor cools and condenses again as a result of the loss of heat and the height and turns from air into water, and having become water falls again onto the earth."

But in most respects Aristotle was dead wrong about the workings of the atmosphere—and vehement in defending all of his ideas. He savaged Anaxagoras, who had suggested—correctly—that hailstorms occur most often in summer, when the heat forces clouds higher into the sky, where the air is colder. "The process is just the opposite of what Anaxagoras says it is," declared Aristotle, insisting that hail "takes place when clouds descend into the warm air." He asserted that "the unscientific views of ordinary people are preferable to scientific theories of this sort."

The power of Aristotle's intellect was such that he not only prevailed in his own time, but created an intellectual hegemony that lasted for nearly 2,000 years. During that long and in many respects dark progression of centuries, little more than folklore was added to man's understanding of the air he breathed and of the complex canopy under which he lived. When the reawakening came, it depended on the very aspect that the Greeks had scorned: on measuring the way things actually work, testing and relentlessly pursuing what is known today as scientific proof.

The man who led the way was the extraordinary Italian Renaissance astronomer and physicist Galileo Galilei, who is best known to posterity by his first name. Endlessly inquisitive, tireless and blessed not only with a stupendous intellect but with supreme confidence in his own abilities, Galileo earned a reputation that shines forth even in comparison with the other superminds of his day. His father, a Florentine musician of modest means,

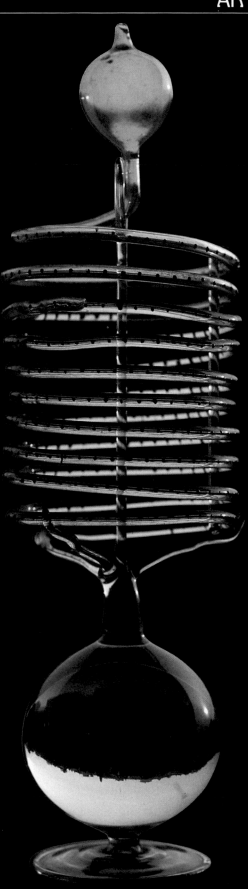

It is no accident that scientific understanding of the atmosphere languished until well into the modern era; without reliable instruments to measure the myriad changes in temperature, humidity and pressure, theories about what caused the variations could not be verified. In the 17th and 18th Centuries, the kinds of tools that were needed finally began to appear. An array of sensitive devices, including the thermometers, barometers and hygrometers (humidity gauges) pictured here and on the following pages, paved the way not only for systematic meteorological observations but for a host of experiments that defined the basic properties of air.

A number of these early instruments originated in Florence, where Grand Duke Ferdinand II and his brother Leopold carried on the Medici family tradition of scientific patronage. The two brought together leading scientists and the region's gifted artisans, who proved no less vital to the task at hand: The delicate, serpentine structure of the spiral thermometer *(left)*, which was designed in 1657, owed as much to the art of fine glass blowing as to the scientific legacy of Galileo.

These efforts were not without their flaws. For one thing, early inventors failed to develop and apply consistent standards of measurement. The first thermometers, for instance, used different temperature scales—greatly complicating the exchange of experimental findings. In time, measurement systems became uniform, to the benefit of the scientific community. But for sheer ingenuity, the pioneer instruments have rarely been excelled.

A 17th Century Florentine spiral thermometer —photographed here with a Medici villa reflected in its base—concentrates nearly nine feet of glass tubing within its 13-inch-high frame. Dots painted on the stem mark degrees.

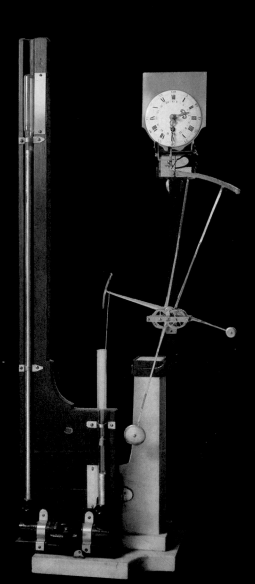

This 18th Century Italian barometer is
operated by a small float that rises and falls as
the column of mercury *(center)* responds
to changes in atmospheric pressure.

This modern replica of a classically decorated
17th Century Florentine hygrometer has a
hollow core that can be filled with ice; moisture
in the surrounding air condenses on the
glass cone and drips into the measure below,
indicating the degree of humidity.

Englishman Daniel Quare's 18th Century
barometer uses a glass vacuum tube inverted in
a cistern of mercury at the bottom of the
stand. The mercury in the tube rises or falls
with changes in the atmospheric pressure.

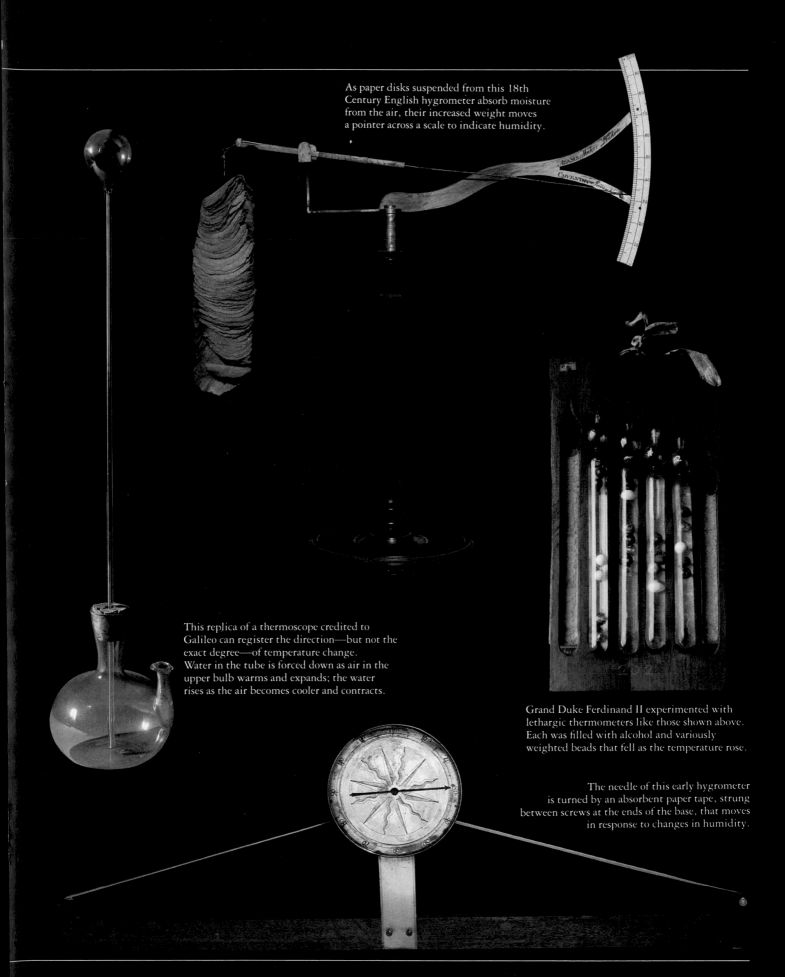

As paper disks suspended from this 18th Century English hygrometer absorb moisture from the air, their increased weight moves a pointer across a scale to indicate humidity.

This replica of a thermoscope credited to Galileo can register the direction—but not the exact degree—of temperature change. Water in the tube is forced down as air in the upper bulb warms and expands; the water rises as the air becomes cooler and contracts.

Grand Duke Ferdinand II experimented with lethargic thermometers like those shown above. Each was filled with alcohol and variously weighted beads that fell as the temperature rose.

The needle of this early hygrometer is turned by an absorbent paper tape, strung between screws at the ends of the base, that moves in response to changes in humidity.

had hoped Galileo would be a doctor, as physicians earned handsome fees. But while studying medicine at the University of Pisa, the young man happened to attend a lecture on geometry. Fascinated, he persuaded his father to let him switch to mathematics and science; the world has been the beneficiary ever since.

Typical of Galileo's intuitive brilliance was a discovery he made one day while sitting in church, observing a chandelier being swung in different arcs by a shifting breeze. By timing the swings with the beat of his pulse, he determined that no matter how narrow or wide the arc, the chandelier completed it in the same amount of time. Later he tested his observation experimentally; he rigged two identical pendulums with bullets on strings, set them swinging through different arcs and confirmed that each oscillation consumed the same amount of time. He had discovered the property of the pendulum, which others would use to make the first truly accurate clocks.

Such insights punctuated Galileo's life. Learning that a Dutchman had invented the telescope, he quickly constructed one of his own. He turned it on the heavens, and found that the Milky Way is composed of innumerable separate stars, saw that the moon has mountains and estimated their height from their shadows, ascertained that Jupiter has four moons of its own, identified sunspots on the sun's surface and used them to prove that the sun rotates on its axis once every 27 days.

Galileo's various investigations frequently brought him into confrontation with the mystery of the nature of air, particularly in his studies of free-falling objects. While it is not true that he tested the behavior of falling bodies by dropping cannon balls of different weights off the leaning tower in his hometown of Pisa—someone else did that—Galileo did study acceleration. The investigation was part of a relentless, almost obsessive, assault on Aristotelian physics.

Aristotle had maintained that an object's speed varied according to its weight (a five-pound rock supposedly fell five times as fast as a one-pound rock, and so on). Galileo demolished this theory, proving mathematically and confirming experimentally that, in principle, all objects fall at the same velocity regardless of weight. To explain why a feather nevertheless floats gently to earth while a rock plummets, Galileo invoked the concept that air itself offers resistance to the fall.

Perhaps intrigued by this notion, Galileo moved on to consider other properties of air. He devised an experiment to determine whether its ability to resist motion meant that it had measurable substance. He fitted a large, narrow-necked glass flask with a valve "through which by means of a syringe," he wrote, "I forced into the flask a great quantity of air." He carefully weighed the flask, which now contained "two or three additional volumes beyond what is naturally contained," then released the compressed air and weighed the flask again. From such experiments he determined that air does indeed have weight; according to his calculations it was 460 times lighter than water. (According to modern analyses, at sea level it is approximately 816 times lighter.)

Aristotle had been typically inflexible on another question about air: The existence of a vacuum, a space containing no air, was impossible, he had ordained, because nature abhorred a vacuum. Just as typically, Galileo took the opposite view. He cited an experiment in which two flat, polished glass plates, when placed together, stuck so tightly that they could be separated only by sliding one across the other. To him, this demonstrated that attempts to pull the plates apart created a vacuum between them and the vacuum held them together. But orthodox physicists, interpreting the same experiment according to Aristotle's precept, replied that nature forced the plates together to prevent a vacuum from forming.

Galileo responded by assembling a device that pulled an airtight wooden stopper away from water in a thick-walled glass cylinder. There, plainly visible, between the stopper and the water was a space that had to be a vacuum. Another nail had been driven in the coffin of Aristotelian theory.

The incontrovertible existence of a vacuum—reports of which were simply not believed by most Europeans—prompted another discovery about the nature of air. Unfortunately, Galileo did not live to achieve it. He knew from pump mechanics that the simple suction pumps of the day—which worked presumably because nature, abhorring a vacuum, pulled the water up to prevent one from forming—could not under any circumstances raise well water more than 34 feet. Galileo is said to have remarked wryly that "nature does not seem to abhor a vacuum above 34 feet," but he offered no satisfactory explanation. He, of course, had discarded the Aristotelian explanation that nature required the water to rise to prevent a vacuum from forming; he knew that a vacuum could exist. But there was one basic flaw in Galileo's thinking about the vacuum—he thought it exerted a pulling force that, in effect, sucked the water upward. He thought that perhaps the force of the vacuum was somehow limited to 34 feet, or that the weight of the water in a column that long simply broke it off.

In late 1641, he apparently posed the problem of the 34-foot limit to a new assistant, 33-year-old Evangelista Torricelli, already a brilliant mathematician and physicist in his own right. Torricelli had worked for 14 years with two of Galileo's foremost disciples, and on their recommendation had been engaged to assist the master himself with a new scientific treatise. But barely three months after Torricelli arrived at Galileo's Florentine villa, their promising scientific partnership was cut short by Galileo's death at the age of 77. As Torricelli prepared to leave the Tuscan court, Grand Duke Ferdinand II named him to succeed Galileo as court mathematician and philosopher.

Torricelli continued to ponder the problem of the 34-foot water column. To simplify the cumbersome apparatus needed to study it—a group of Roman physicists had been working with tubes attached to multistory buildings—he recalled Galileo's suggestion that the height of a liquid column would vary in inverse proportion to the liquid's weight, and tried heavier liquids such as sea water, honey and mercury, which decreased the height of the liquid column and allowed him to use much shorter tubes. Finally, in 1644, he had his laboratory assistant, one Vincenzo Viviani, fabricate a three-foot-long glass tube strong enough to hold a column of mercury, a liquid metal that weighs 14 times as much as water. Following Torricelli's instructions, Viviani filled the tube with mercury and plugged its end with his finger, then turned the tube upside down and submerged the open end in a basin of mercury before unplugging it. The mercury column descended until it was about 30 inches tall, leaving a vacuum at the top of the tube. They had invented the barometer.

The true hallmark of Torricelli's genius was not the experiment itself, which was merely a variation on earlier procedure, but its interpretation. The column of liquid was not being pulled up by the vacuum, he concluded, but was being pushed up by the weight of the atmosphere, which bore down on the mercury in the pan but not on the liquid in the column. Torricelli went further, and used his apparatus to "show the changes of the air, now heavier and coarser, now lighter and more subtle." To prove that the column was supported by the weight or pressure of the outside air rather than by Galileo's "force of vacuum," Viviani fabricated a similar tube, this one sealed at one end by a large, hollow bulb, which would create an even larger, and by Galileo's reasoning a stronger, vacuum. When this tube was filled and inverted, the mercury descended to the exact level of the original one—"an almost certain sign that the force was not within," wrote Torricelli. He came to a conclusion of elegant simplicity and lasting significance: "We live submerged at the bottom of an ocean of elementary air, which is known by incontestable experiments to have weight."

But in a key respect Torricelli was utterly frustrated. "I have been unable to find out with the instrument when the air is coarser and heavier and when more subtle and light," he reported to a colleague, "because the level is very sensitive to heat and cold." Historians suspect that the vacuum was contaminated by water vapor, whose reaction to temperature variations would mask pressure changes.

Torricelli described his invention in a letter to a trusted friend but prudently refrained from publicizing it—apparently for fear of antagonizing the theologians of the Inquisition, who 11 years before had forced Galileo to recant his unorthodox opinions about the solar system. Catholic dogma in 1644 remained wedded to Aristotelian doctrine, and the very concept of a vacuum was considered blasphemous. Such constraints were weaker outside Italy, however. Several Frenchmen witnessed Torricelli's occasional demonstrations of his vacuum apparatus, and their accounts soon prompted similar experiments abroad.

Among the scientists intrigued by the new instrument was Blaise Pascal, a famous French prodigy who at the age of 16 had discovered major geometrical theorems and by the age of 21 had invented an expensive but workable mechanical calculator, the earliest precursor of the cash register.

In 1648, at the age of 25, Pascal was thoroughly familiar with Torricelli's proposition and was determined to verify one of its main aspects. If the air was an "ocean," and had weight, then the upper air pressing down would add its weight to the air below, increasing the pressure there. If Torricelli was right, then just as the pressure, or cumulative weight, of water increases with depth, so the pressure of air should decrease with height. The way to find out, Pascal realized, was to carry a barometer to the top of a mountain.

He could not do the climbing himself for a variety of reasons—he was frail and chronically ill, he feared scaling heights, and in any case there were no appreciable mountains near Paris. By chance, however, his brother-in-law, a government functionary named Florin Perier, lived in the mountainous Auvergne region of southern France. Perier agreed to conduct the experiment. Pascal shipped him the materials and on September 19, 1648, Perier proceeded, recording the results in a long written report.

To witness the experiment, he wrote, he had invited a number of distinguished clerics and laymen of his town of Clermont to accompany him.

First they repaired to the garden of the Minim Fathers, a low spot in town, where Perier set up a barometer in the Torricelli manner. Carefully and repeatedly, they measured the height of the mercury—exactly 28 inches. Leaving one of the Minim monks, "a man as pious as he is capable," to stay in the garden with the barometer to see whether the level changed during the day, Perier and his entourage scaled the 4,888-foot Puy de Dôme and set up another barometer at the summit. The reading: 24⅔ inches. "We were so carried away with wonder and delight," reported Perier, "and our surprise was so great, that we wished, for our own satisfaction, to repeat the experiment." He moved the barometer about the summit and took five additional readings. All were identical. As the elated group descended, Perier set up the tube again and found that its reading had increased. Back in the garden, the watchful monk reported that the reading there had not changed. Now the meticulous Perier once more set up the tube he had taken up and down the mountain. In the garden it, too, registered 28 inches, "and thus finally verified the certainty of our results."

The results absolutely confirmed Pascal's belief that air is an elastic substance whose weight, or pressure, decreases with altitude. The proof, he said, afforded him "great satisfaction."

Although much had been revealed about air by Torricelli and Pascal, their attitude toward it as a single, homogeneous substance blinded them to the resolution of any number of persistent mysteries. What, for instance, happened to water when it evaporated? The Greeks had thought it turned into air, but that left the question of how it turned back into water to form clouds. The first productive description of air as something more than a single substance was offered by a French contemporary of Galileo's who had little interest in the atmosphere.

René Descartes had much in common with Aristotle; there are those who regard the solitary Frenchman as the father of modern philosophy, Aristotle the progenitor of Classical thought. As the Greek loved logic, Descartes loved mathematics, "because of the self-evidence of its reasonings." Descartes deeply distrusted Aristotelian philosophy, in which he had been rigorously schooled, yet committed similar errors when he stubbornly tried to impose universal laws on an intractable variety of particular situations. What Descartes had to say about air is an almost incidental footnote in the context of his works and reputation—as indeed was Aristotle's musing on the subject—but remains fascinating because it contained a first, shadowy glimpse of the truth.

In an appendix to his famed *Discourse on Method,* Descartes proposed that all earthly substances were composed of a single fundamental substance; minute particles of this basic material, of varying sizes and shapes, were suspended in an invisible "subtle matter." The fundamental particles of air, he said, were "laid on one another without being interlaced," while in solid substances the particles "intertwined to become hooked and bound to each other, as are the various branches of bushes that grow together in a hedgerow." Water he described as being composed of smooth, slippery particles "like little eels, which join and twist around each other but do not knot or hook together in such a way that they cannot easily be separated."

Furthermore, according to Descartes, water could change from a solid to a liquid or gas, depending on how tightly its particles were hooked together. Evaporation occurred when the subtle matter, "being strongly agitated

A fanciful 1698 engraving depicts a welter of cultural and scientific activities at a meeting of the French Academy of Sciences. While the meetings were in fact far more austere and private, the French Academy and its European counterparts provided frequent and substantial support for atmospheric research.

by the sun or for some other reason," in turn agitated water particles. This caused particles "which are small enough, and which can be separated easily from their neighbors, to break away here and there and fly into the air," just as "the dust of a field flies up when it is merely disturbed and propelled by the feet of a passer-by."

After Descartes had thus hinted that air might be a mixture of constituent gases rather than a single substance, scientists increasingly became engrossed in finding out what the constituents might be. Whereas Galileo and his colleagues had been physicists and astronomers concerned with the mechanics of air flow, the key contributions in the late-17th and 18th Centuries were made by chemists who attempted to break down and analyze the peculiar invisible substance. Of these, the pioneer was the aristocratic, frail and ascetic English experimenter Robert Boyle.

The seventh son of a well-to-do Anglo-Irish landowner, Boyle never cared for the world of property and power, preferring instead to devote himself to scientific exploration in the laboratory. At the age of 14 he had avidly consumed the works of Galileo, and in his twenties, without ever having attended a university, he rented a suite of rooms in the town of Oxford and began what was to be an impressive string of inquiries and discoveries.

He had learned of an air pump developed by a German physicist, and he decided to build a better one in order to pursue research on air and combustion. He was fortunate in having a spectacularly talented young physicist named Robert Hooke as an assistant. In fact, Hooke was destined to become almost as famous as Boyle, but more as an inventor than a theoretician; among other things, during his long, productive but disputatious life, he invented the first practical anemometer for measuring wind speed, the universal joint used today on automobile drive shafts, the iris diaphragm still used in most cameras and the spirit level used in construction.

All this lay ahead for Hooke, however. With Boyle he exhaustively ana-

Spherical particles of vapor rise from a body of water in this woodcut from René Descartes's *Discourse on Method*. The French philosopher theorized that water droplets agitated by the sun's rays separate and rise into the air —a rudimentary explanation of evaporation.

A portrait of the 17th Century British chemist Robert Boyle portrays the elegant aristocrat reading at leisure. Although he took up chemistry as an avocation, he proved to be one of its most important pioneers.

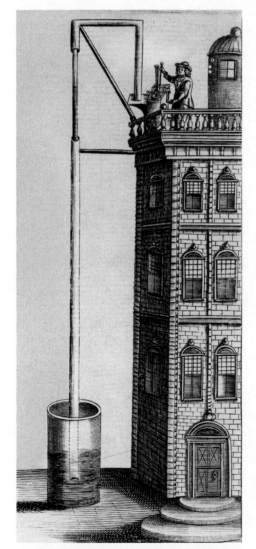

From a lofty rooftop, Robert Boyle prepares to observe the effects of atmospheric pressure by pumping the air out of a pipe set into a cistern of water. The cumbersome apparatus was necessary because air pressure raised the water in the pipe some 34 feet.

lyzed the German air-pump design and in 1657 came up with a greatly superior device. Boyle was able to produce an almost total vacuum with it, and one of his first experiments was to test Galileo's notion that a feather and a heavier object, in this case a coin, would drop at the same rate in an airless enclosure. Yes, he found, they did reach the bottom at virtually the same instant.

With his pump, Boyle was able not only to remove air from a space, but to force air into it. After compressing a large quantity of air into a glass chamber and sealing it, Boyle found that even after a long period of time the pressure within the chamber remained undiminished. He concluded that "there is a spring or elastic power in the air we live in," and speculated that it contains coiled particles that expand as soon as a confining force is removed. To analyze this phenomenon, he and Hooke combined a mercury barometer with a network of sealed glass tubes and simultaneously measured both the pressure and the volume of the trapped air. From a series of 44 meticulous measurements, they derived a simple mathematical relationship that became known as Boyle's law: If the temperature remains unchanged, the volume of a parcel of gas varies inversely with its pressure—doubling the pressure halves the volume, and vice versa.

Boyle and Hooke also experimented with combustion and respiration. They took to enclosing mice, birds and other creatures in glass jars and observing them. If confined for an extended period, an animal would begin to pant and weaken, but when released from the jar it would quickly revive. If the air was removed with the pump, on the other hand, the animal would suffer immediately and its life would soon be endangered. Clearly, air was necessary for life—and it contained something that animals used up in their breathing. The same held true for combustion. A lighted candle placed in the enclosed jar would go out as the air was evacuated; when the air was compressed, the flame burned longer than it did in a sealed, unpressurized chamber. But Boyle had no idea what the mysterious, life-giving substance was. One day it would be known as oxygen.

The promise of Boyle's work was not realized for more than a century, partially because of one of the more celebrated blind alleys in the history of science. About 1700, chemistry came to be dominated by the concept of phlogiston, a hypothetical material that supposedly was released into the air whenever a substance burned or rusted. Phlogiston was thought capable of flowing from one substance to another, and of extinguishing a flame when present in too high a concentration (thus the candle burning in the sealed chamber eventually went out). This concept had a major logical flaw: The weight of phlogiston had to be regarded as either positive or negative, because rusting substances gain weight, whereas burning substances lose weight. But since the concept explained many chemical results while retaining aspects of the Aristotelian system, phlogiston achieved wide acceptance and swirled like a fog in the thinking of the ensuing pioneer chemists.

Nevertheless, as 18th Century chemists continued to analyze solids and liquids and to identify their constituent parts, air inevitably came under scrutiny. The first of its components to be identified was a minor one, but it was vital, and its isolation led to the analysis of others.

The discoverer was a Scottish medical student named Joseph Black, who in 1753 happened to be investigating the medicinal properties of chemicals

27

derived from limestone (calcium carbonate, in modern chemical terms) and *magnesia alba* (magnesium carbonate). When limestone was treated with acid, Black discovered, it gave off a strange kind of "air" that would instantly extinguish a burning piece of paper, "as if it had been dipped in water," and would suffocate a small animal. Because this mysterious air could be absorbed, or "fixed," by derivatives of lime or magnesium, he named it "fixed air"; today the gas is known as carbon dioxide. As Black pursued his investigation of this fascinating stuff, he found that fixed air not only was present in solid materials such as *magnesia alba,* but was "dispersed thro' the atmosphere," and was particularly evident in expelled human breath and the bubbles given off by fermenting beer (which he analyzed at a nearby brewery).

It was now known that the atmosphere was far more complex than previously suspected; it contained Black's fixed air, which extinguished flames and life, and Boyle's "vital substance," which intensified and prolonged burning. But when the vital air had been burned or breathed out of a sealed chamber, and the fixed air had been absorbed, the great majority of the sample of atmosphere remained. What was in it?

The question was still on Black's mind almost two decades later, by which time he had become a professor of chemistry at the University of Edinburgh. He was much too busy to investigate this problem further, but it happened that one of his best students, a medical student named Daniel Rutherford, was in 1770 looking for a subject for his dissertation; Rutherford enthusiastically took on the problem of isolating what Black referred to as residual air. Removing the fixed air (carbon dioxide) by absorbing it was easy. The problem was getting the vital air (oxygen) out, because whatever consumed that substance, whether living creature

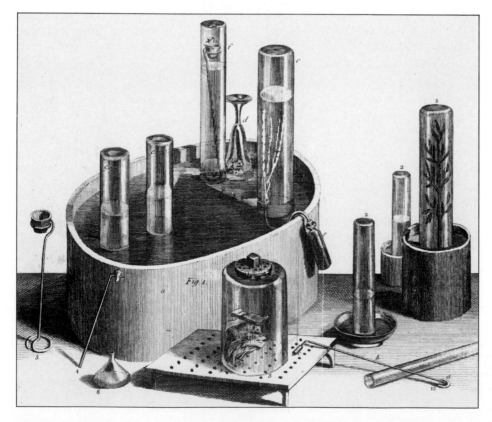

In the late 18th Century, Joseph Priestley placed mice, plants and other specimens in gas-filled flasks inverted in water to determine the effects of different "airs," or gases. Among other discoveries, he found that a mouse became animated in "dephlogisticated air" (pure oxygen) and a plant *(right)* could restore oxygen to a container that had been deprived of it.

French chemist Antoine Lavoisier supervises a 1789 experiment in which a subject breathes pure oxygen while pumping a foot treadle. The simple trial—recorded by Madame Lavoisier, who included herself in the picture—showed that exercise increases consumption of oxygen.

or flame, simultaneously gave off fixed air, as Black had demonstrated.

After several false starts, Rutherford devised a way to do it. He captured a sample of the atmosphere by inverting a jar in a pan of water. After absorbing the fixed air in the jar with caustic potash, he burned phosphorus in the remaining air. Phosphorus, he had found, totally consumed the vital air, and instead of generating fixed air gave off fumes that were absorbed by the water in the pan. The air that remained—about three quarters the volume of the original sample—instantly extinguished a flame or life, and Rutherford labeled this third and, it was thought, last component of the atmosphere "noxious air." Although other researchers produced similar results at about the same time, Rutherford is generally credited with the discovery of nitrogen, as it was later named—despite the fact that he persisted in trying to explain its properties in terms of phlogiston and thus failed to understand his own achievement.

Indeed, the same thing was true of the other chemical pioneers who had unveiled the basic constituents of air; they had so far understood only dimly the properties and the importance of the substances they had observed. The latter part of the 18th Century was a period of extraordinary experimental activity during which scores of scientists made incremental contributions to knowledge about air, while a few stumbled upon insights of profound significance.

One of these few was Joseph Priestley, a nonconformist English minister whose radical ideas included outspoken support for the rebellious American colonists. He was also a self-taught chemist who became engrossed in the characteristics of fixed air. In 1772, he managed to dissolve some of it in water, tasted the result and found it tart and refreshing; he had created what eventually came to be known as soda water. His researches, along with his political and religious extremism, impressed a wealthy nobleman, Lord Shelburne, who offered Priestley a position as personal librarian and resident intellectual. Priestley happily accepted this subsidy for his intellectual pursuits, and it was while working for Shelburne that he made his most notable contribution.

When Priestley heated some mercury in open air, he found that it formed a brick-red calx now known as an oxide of mercury. Then, when he went on to heat this calx in a test tube, it gave off a gas that possessed truly remarkable qualities. He noted that he felt "light and easy" when he breathed this gas, and mice became positively frisky in it. Priestley already knew that a mouse confined in a bell jar containing two ounces of ordinary air would expire in 15 minutes; he discovered that a mouse confined in the same amount of this new substance would emerge from the jar after half an hour suffering only a slight chill.

He noticed something else of great significance. The ordinary air that had been rendered unfit by the breathing of a human or animal (Black's fixed air) would cause a plant to luxuriate. "In no other circumstances," he said, "have I ever seen vegetation so vigorous as in this kind of air, which is immediately fatal to animal life. This observation led me to conclude that plants, instead of affecting the air in the same manner with animal respiration, reverse the effects of breathing, and tend to keep the atmosphere sweet and wholesome, when it is become noxious, in consequence of animals either living and breathing or dying and putrefying in it."

Still under the spell of controversial scientific wisdom, Priestley tortuously explained the behavior of his "sweet" and "noxious" gases in terms of phlogiston, introducing such unwieldy terms as "dephlogisticated" air. But in 1774, during a trip to the Continent with Lord Shelburne, Priestley expounded on his new gas to an intrigued dinner-party audience that included the brilliant French chemist Antoine Lavoisier. The circumstantial conversation was to be the death of phlogiston.

Lavoisier had long been skeptical about the existence of phlogiston. A prodigious worker who earned a law degree while training to become a scientist, Lavoisier continually involved himself both in public affairs and with various practical challenges such as the need for improved street lighting in Paris and the modernization of agriculture. His investigation of street lighting was what got him interested in combustion and the behavior of heated metals.

During the 1770s, Lavoisier, like Priestley, had been investigating an odd thing that happened to most metals when they were heated. A powdery substance formed on their surfaces, and because it usually looked something like lime, it was named calx, the Latin for "lime." Heating various metals in sealed containers to produce the calx, Lavoisier noted that as it formed, there was no change in the total weight of the sample and its container; but when he opened the container to the air, there was a slight increase in weight. He deduced that some air had been consumed by the metal in the formation of the calx.

Now came Priestley to announce that by heating a calx he had generated an extraordinary kind of air. Lavoisier promptly repeated Priestley's experiment, and he made much more of it than Priestley had been able to. Some

While attempting to make an improved telescope lens in the early 19th Century, German optician Joseph Fraunhofer inadvertently created a device that would one day transform atmospheric research. Called a spectroscope, his invention employed a prism and an arrangement of lenses to cull discrete wavelengths from the full spectrum of sunlight.

Though his prototype apparatus was relatively crude, Fraunhofer discovered that the solar spectrum is riddled with fine, dark lines, which have come to be known as Fraunhofer lines. He also found distinctive dark-line patterns in starlight and curious bright-light spectral patterns in candlelight.

The strange patterns remained a puzzle to scientists until 1859, when chemist Robert Bunsen and physicist Gustav Kirchhoff began experiments with their own version of the spectroscope. Analyzing the light produced by the combustion of different chemicals (using the now-famous Bunsen burner), they discovered that, when burned, each element emits light in a characteristic and unique pattern. Kirchhoff noticed that some dark Fraunhofer-line patterns matched the bright-line patterns, and from this he concluded—correctly— that an unburned element absorbs the same light wavelengths that it emits when burned. Thus, the chemical makeup of any substance can be revealed both by the way it creates light when burned and by the way it blocks, or absorbs, light from an outside source.

With this new tool, scientists were able to identify previously unsuspected trace elements in the atmosphere, such as argon and helium.

Reading the Language of Light

This spectroscope, built in 1860, enabled Robert Bunsen and Gustav Kirchhoff to discover new elements. Bunsen and Kirchhoff's techniques were used by later scientists to identify the components of the atmosphere.

In this diagram of the workings of a spectroscope, light from a burning substance enters at right, passes through a lens, and is dispersed into spectral colors by the prism. A second lens projects the spectrum onto a focal surface for viewing through the tube at left. The short third tube superimposes a gradient on the spectrum for analysis. The spectra of cesium, the first element thus identified, and of ordinary sunlight are shown at the bottom.

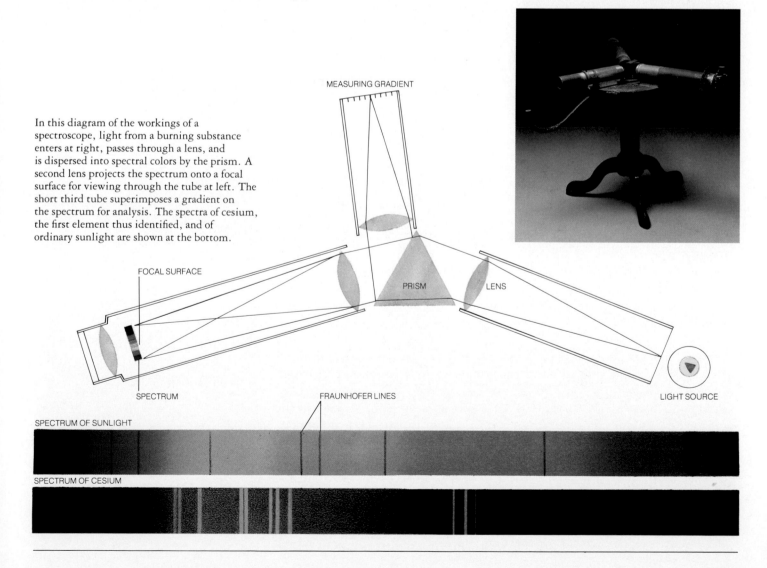

MEASURING GRADIENT

FOCAL SURFACE

PRISM

LENS

SPECTRUM

FRAUNHOFER LINES

LIGHT SOURCE

SPECTRUM OF SUNLIGHT

SPECTRUM OF CESIUM

"more salubrious and purer portion" of the air, he later said, combined with metals to form calx and was regenerated by heating the calx of mercury. Air appeared to be one part pure air, which supports burning and respiration, and three or four parts "deleterious" air, which supports neither. The pure air Lavoisier named oxygen—ironically, for the wrong reason. He mistakenly thought the intriguing gas was the basis of all acids and gave it a name derived from the Greek for "acid producer."

Before very long, Lavoisier had come to the conclusion that fixed air was composed of carbon and oxygen—an analysis that eventually led to its being renamed carbon dioxide. It was, he knew, present throughout the atmosphere, but in quantities too small to be measured by the equipment available in the 18th Century. Air's principal components, according to Lavoisier's calculations, were oxygen (between one quarter and one third by volume) and deleterious air, renamed nitrogen a few years afterward by another researcher.

In 1783, Lavoisier assembled the lessons of his experiments in a paper titled "Reflections on Phlogiston." He reemphasized that the formation of calx, now known to be oxidation, involved no change of weight in a sealed chamber; no mysterious extra "principle" was needed to account for this kind of combustion. "If everything is explained in chemistry in a satisfactory manner without the aid of phlogiston," he observed, then it is "infinitely probable that this principle does not exist."

Sadly, both Priestley and Lavoisier fell prey to political misfortune during the French Revolution. The freethinking Priestley enthusiastically espoused the Revolution and criticized political discrimination against theological dissent, thereby making powerful enemies in conservative England. Eventually, a mob pillaged his house and wrecked his scientific apparatus, and Priestley later fled to the United States, where he spent the last 10 years of his life. Lonely and isolated from world affairs at his home in Pennsylvania, Priestley nevertheless became a friend of President Thomas Jefferson, and professed to find himself for the first time in his life in a friendly political environment.

Lavoisier's association with a prerevolutionary tax-collecting body incurred the fury of French extremists. During the Reign of Terror the great chemist was arrested, tried on trumped-up charges and executed at the guillotine. His epitaph might be a lament from a fellow scientist: "It took them only an instant to cut off that head, and a hundred years may not produce another like it."

The analysis that was sketched out by Lavoisier and refined by him and others—that 21 per cent of the atmosphere was oxygen and 79 per cent nitrogen, with just a smattering of carbon dioxide—was complete for all practical purposes, and it persisted for a century. In 1892, however, the eminent professor of natural philosophy at the Royal Institution in London, John William Strutt, Lord Rayleigh, found that pure nitrogen manufactured from ammonia weighed slightly less than nitrogen obtained from the air. The discrepancy between the two was only .5 per cent, but that was enough to intrigue him. It happened that a fellow chemist, William Ramsay, heard of Rayleigh's anomaly, and independently took up the pursuit of the elusive remnant.

In 1894, each of them succeeded in isolating a trifling amount of the

Lord Rayleigh (*above*) found in the 1890s that air is composed of more than nitrogen, oxygen and carbon dioxide. The ensuing isolation of argon helped earn him and colleague William Ramsay (*below*) Nobel Prizes.

Sir William Ramsay points proudly to a section of the periodic table of elements in this 1908 illustration. The gases in the right-hand column—neon, argon, krypton and xenon —were isolated by Ramsay and his co-workers.

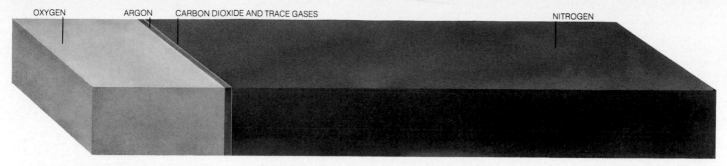

As illustrated above, nitrogen and oxygen make up the lion's share of the earth's atmosphere, averaging 78 per cent and 21 per cent of the total respectively. Argon amounts to just over .9 per cent by volume, and the minute remainder is left to carbon dioxide and traces of scores of other gases. The percentage of water vapor (not shown) fluctuates markedly.

unknown gas. Unable to identify the substance chemically, they decided to experiment with a relatively new technique called spectroscopy. It had been known since Isaac Newton's time that light passing through a prism separates into a banded spectrum of all its colors. In 1853, physicists had made the discovery that light passing through different gases produces different spectra, and that they could identify the gas by the spectrum thus produced.

Within a few days of each other, Rayleigh and Ramsay trained spectroscopes on their unknown gas, caused the sample to glow by passing an electric spark through it, and found in the resulting spectrum a blue line produced by no other element. What they had was a new element, which they named argon; it proved to make up just under 1 per cent of atmospheric air. During the following year Ramsay, while trying to find argon in one of the earth's minerals, isolated a gas that confounded him by revealing yet another kind of spectrum, this time with a bright yellow line. The gas was helium, hitherto detected only on the surface of the sun—again by spectroscopy—but now traceable in the atmosphere. Before he was finished, Ramsay, in collaboration with chemist Morris William Travers, had isolated three more gases that are found, if only in minute quantities, in the air: krypton, neon and xenon. All of them had been differentiated by means of the spectroscope.

Thus the tenuous and elusive air had at last yielded to complete analysis; but it was hardly a final victory. These secrets had been wrested from readily available samples of air found literally under the very noses of researchers. Far beyond the reach of 19th Century technology, at the edge of space and in the farthest reaches of the globe, there still lurked magnificent mysteries of the atmosphere's behavior and structure. Ω

FIERY FANTASIAS ALOFT

The journey of sunlight through the atmosphere's richly varied mix of gases and particles produces a never-ending global light show. Some of the optical effects are so common they are scarcely noticed: For example, dust and gas molecules scatter those wavelengths of light perceived as blue, creating the blue of the daytime sky. Other optical phenomena occur only in special circumstances. The majestic mock suns pictured at right, for instance, are caused by the refraction (bending) of light by hexagonal crystals of ice in the air.

Smaller, prism-like crystals can fashion bright halos around the sun or moon. Larger crystals sometimes reflect and disperse sunlight to project the vertical beams known as sun pillars.

When a thin cloud of water droplets moves in front of the sun, it may occasion a more complex light phenomenon called diffraction, spawning the gaily hued bands of coronas or glories. Light waves curve around the droplets much as a wave of water washes around an obstruction. The disturbed light waves interfere with one another to yield bands of various wavelengths, perceived as different colors.

Another category of optical marvels depends on the air's density rather than on its contents. For most of the day, the appearance of the sun is not noticeably affected by variations in the density of the air; but at sunrise and sunset, when light waves reach an observer only after traveling a long, oblique path through the dense lower atmosphere, the shape of the sun may appear fantastically altered.

A vibrant circumhorizontal arc tints wispy cirrus clouds above a California hillside. The effect appears when the sun is high in the sky and its rays are bent and dispersed by ice crystals.

A full halo rings the moon in a photograph taken off Tierra del Fuego. Halos testify to the presence high in the atmosphere of ice crystals, which refract incoming rays of light.

Like the path of light often reflected from water by the setting sun, a pillar probes the sky near Nashville, Tennessee. In this case, airborne ice crystals act as mirrors and reflect the light.

The icy air above a station in Antarctica affords a concert of optical effects: A solar halo broadens toward the horizon to reveal mock suns, while a sun pillar bisects the arc.

Sunlight curving around the water droplets
in thin clouds endows the sun with a delicately
tinted corona *(left)*. The resulting chromatic
patterns, which are extremely diffuse,
are detailed in the telephoto view below.

A colorful aura known as a glory tinges a
fog bank around the shadow of a photographer
who is standing on a peak in Japan *(top)*.
Often seen surrounding an airplane's shadow
(above), the effect occurs when cloud
droplets reflect and scatter light waves.

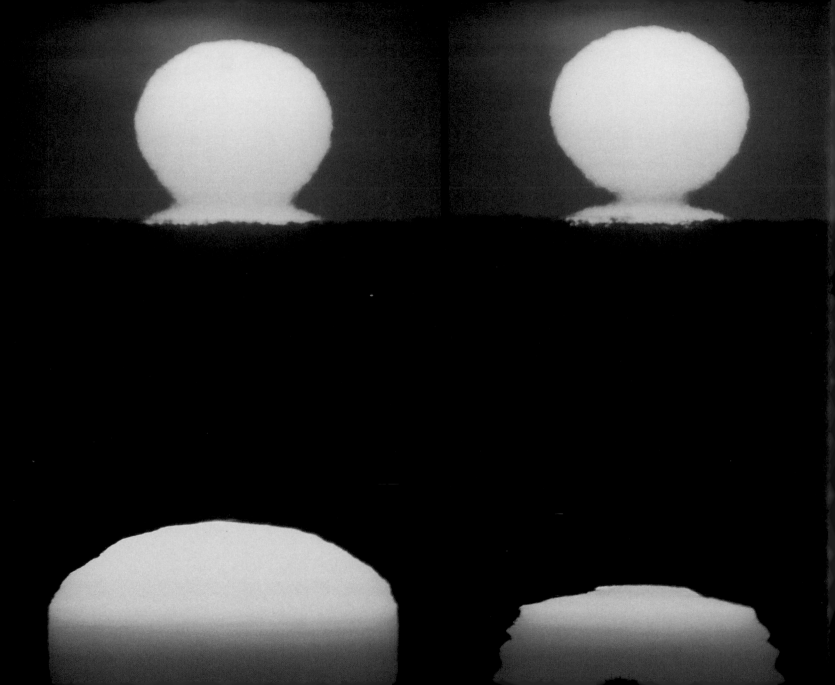

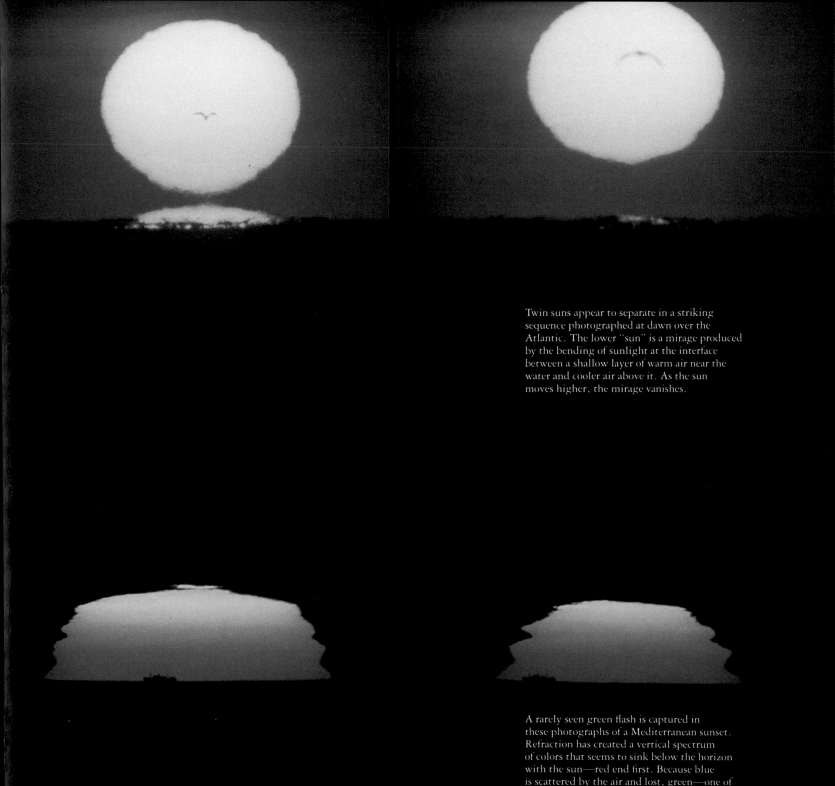

Twin suns appear to separate in a striking sequence photographed at dawn over the Atlantic. The lower "sun" is a mirage produced by the bending of sunlight at the interface between a shallow layer of warm air near the water and cooler air above it. As the sun moves higher, the mirage vanishes.

A rarely seen green flash is captured in these photographs of a Mediterranean sunset. Refraction has created a vertical spectrum of colors that seems to sink below the horizon with the sun—red end first. Because blue is scattered by the air and lost, green—one of

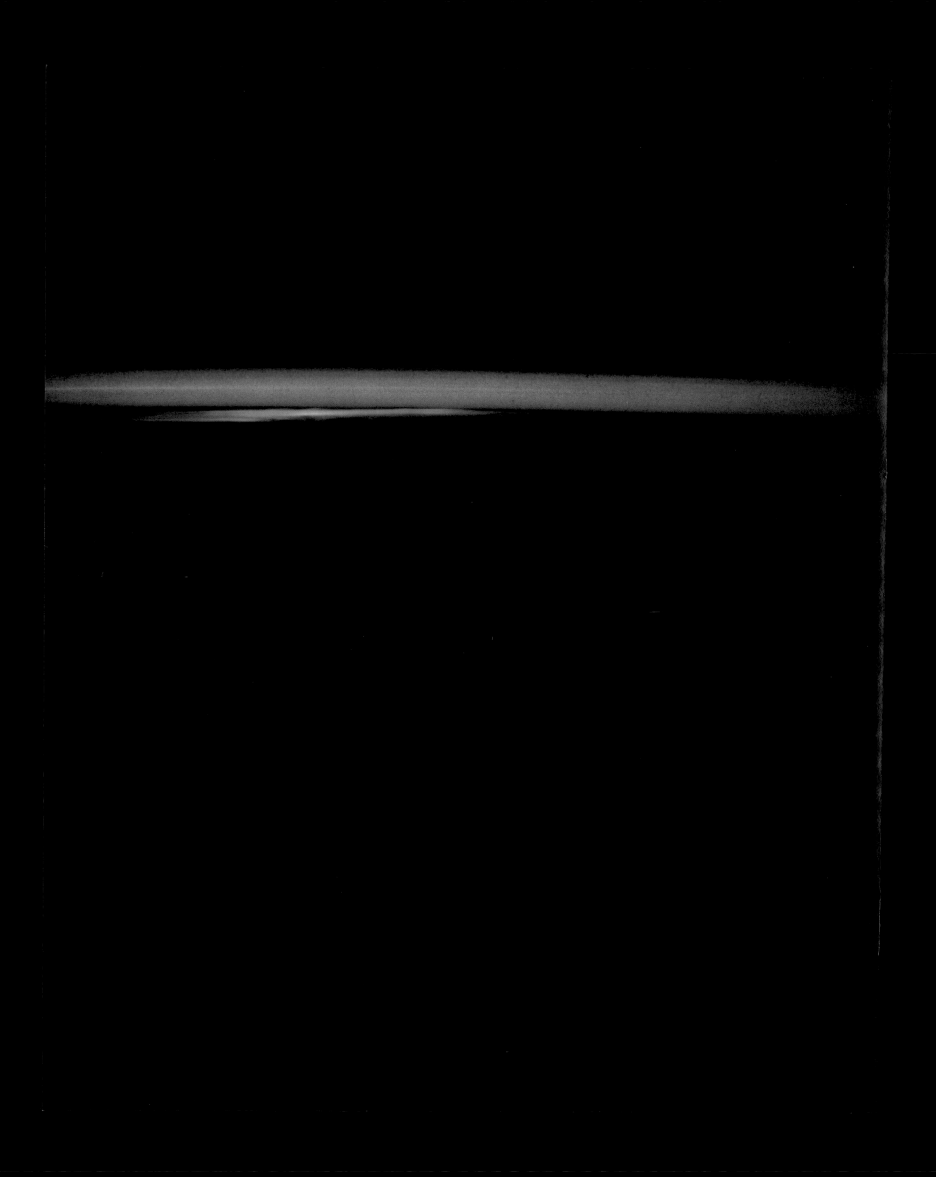

Chapter 2

PROBING THE STRUCTURE OF THE SKY

As Paris sweltered through the summer of 1899, the city's residents, unusually insensible to the wretched heat, delayed their customary August exodus to cooler climes. The entire capital was preoccupied with the retrial of Army captain Alfred Dreyfus, the source of a controversy that threatened the very survival of the Third Republic. Five years earlier, Dreyfus, a Jew, had been court-martialed as a German spy and sentenced to life imprisonment on Devil's Island, but evidence had emerged later that the Army command had framed him with forgeries and perjured testimony. Dreyfus had become the focal point of a vicious political battle between French conservatives—monarchists, clergymen, Army officers and anti-Semites— and the republicans and socialists of the left. Because the volatile proceedings had been removed to Rennes, a Breton city about 170 miles distant, the Paris newspapers' only source of information about the scandalous revelations of the month-long trial was the telegraph. By September 9, with the verdict expected at any moment, interest had reached fever pitch. Suddenly the telegraph line between Paris and Rennes went dead—a result, it developed, not of political conspiracy but of atmospheric research.

That morning, a wealthy French meteorologist named Léon Philippe Teisserenc de Bort had routinely launched a string of box kites from his private observatory in Trappes, 17 miles southwest of Paris. The uppermost kite carried a payload of instruments, driven by clockwork, that continuously recorded temperature, humidity and barometric pressure. Ten other kites spaced at intervals along the steel piano-wire tether helped to support its weight. With Teisserenc de Bort and his assistants steadily unreeling the wire, the leading kite had ascended nearly two miles; when they began winching the kites back down, the wire suddenly snapped. The kites slowly drifted downward over a suburb of Paris, draping it with four miles of wire and wreaking havoc on the ground. The wire fouled the propeller of a Seine steamboat, nearly garroted a policeman and silenced the Rennes telegraph.

While the debacle did not long delay the news of Dreyfus' fate (he was convicted again, but was soon pardoned and eventually awarded the Legion of Honor), it aroused considerable public rancor. And in similar mishaps during the ensuing years, falling piano wire short-circuited Paris' proliferating electric-power network so frequently that Teisserenc de Bort was forced to abandon his kite experiments altogether. The setback did not diminish his enthusiasm for meteorological research; he was determined to probe above the clouds to learn more about the atmosphere—how thick it is, and how its heat, pressure, moisture and gases are distributed at higher alti-

Sunlight from below the horizon, recorded in a photograph taken from an orbiting spacecraft, reveals blue-tinted layers—bands of aerosols of different densities—in the stratosphere.

tudes. He could not have known that he was opening what would become one of the most far-reaching lines of inquiry of 20th Century science.

The investigators who followed him would soon perceive within the atmosphere a subtle architecture—a progression of onion-like layers, defined by various characteristics such as temperature trends, density, electromagnetic properties, and concentrations of gases, radiation and suspended particles. For decades to come, the exploration of each successive spherical envelope and the understanding of each set of relationships would suggest more questions, until the investigation extended all the way to the sun.

Like most pioneers, Teisserenc de Bort was denied the ultimate vista; his role was to take the first step into the strange new world and report on what he saw. As a frail, sickly child, son of a rich and famous government minister, he had passed much of his youth in the salubrious Mediterranean air of Cannes and Grasse; beneath their azure skies his passion for atmospheric science was conceived. At the age of 18 he organized a complete observatory in Grasse and seven years later joined the Paris staff of the infant Central Meteorological Bureau. He rose rapidly to the rank of chief meteorologist, but the government's paltry research budget so frustrated him that he resigned in 1892—the same year that a French aeronaut named Gustave Hermite first launched a barometer aboard an unmanned hydrogen balloon.

Reduced to the status of dedicated amateur, Teisserenc de Bort nevertheless continued to devote his private fortune and all his energy to atmospheric science (serving for many years as secretary general of the French Meteorological Society), and Hermite's balloon experiment especially fired his imagination. Several aeronauts had died as they rose into the cold, thin air above 25,000 feet, but Hermite had shown that simple, inexpensive balloons could carry instruments beyond the limits of manned flight. In 1896, Teisserenc de Bort founded the private observatory on his farmland in Trappes where he conducted his notorious kite experiments and refined the design of his balloons and instruments.

Hermite had used balloons made from silk or the intestinal membranes of cattle, but Teisserenc de Bort settled on varnished paper balloons because they could be designed to burst at a predictable altitude. He soon developed a method of ferrying each instrument package aloft beneath two balloons: a guide balloon containing just enough hydrogen to support the instruments; and a fully inflated "sounding" balloon—a term borrowed from mariners—that actually lifted the apparatus. As atmospheric pressure decreased during the ascent, the hydrogen steadily expanded until it burst the sounding balloon at an altitude that could be manipulated by changing the amount of hydrogen used or the size of the balloon. The weight of the sounding balloon's remains, combined with that of the instruments, dragged the guide balloon slowly back to earth.

To gain accurate wind and altitude measurements, Teisserenc de Bort's assistants tracked the balloons with two theodolites—surveyor's instruments that measure vertical and horizontal angles with the aid of a telescope. Synchronized sightings, taken every 60 seconds, allowed the observers to calculate the balloon's course by triangulation until it disappeared into the clouds. From this data they could decide what part of the countryside to search for the fallen instruments. At sea the balloons were tracked with a sextant, and retrieval was considerably easier because the broken sounding balloon acted as a sea anchor while the red-banded guide balloon signaled the instruments' location to the trailing research vessel.

Teisserenc de Bort also designed a miniaturized, one-pound instrument package for his sounding balloon: It contained a barometer to measure pressure (and hence altitude), a hygrometer to measure the air's moisture content, and a thermometer. Because ink freezes at high altitudes, where temperatures may be as low as −100° F., the instruments operated sharp metal styluses, which traced across a slowly revolving smoke-blackened aluminum drum driven by clockwork; at sea the tracings were etched on copper with pens using sulfuric acid so that the result could not be washed away.

Ordinary thermometers posed other problems; since they responded to temperature changes slowly, their readings tended to lag behind the true temperature as the balloon ascended, and they were susceptible to direct warming by the sun's radiation. To surmount these difficulties, Teisserenc de Bort fabricated a tiny bimetal thermometer from strips of copper and steel and mounted it so that it projected outside the cork instrument box on a nonconducting ebonite strip, where wind could thoroughly ventilate it. During daylight hours the thermometer was shielded by a foil sunshade, and as an additional safeguard some soundings were conducted at night.

The information Teisserenc de Bort derived from his ingenious instruments and techniques defied common sense. For centuries it had been assumed that temperature steadily decreased with altitude until the atmosphere merged with the frigid void of interplanetary space. A few readings taken by unmanned balloons had contradicted conventional wisdom, but most scientists ascribed them to faulty instruments. Teisserenc de Bort was seeing evidence that the old assumption was wrong and, rather than dismissing it, he doggedly pursued the anomaly with repeated soundings.

On April 28, 1902, he presented to the French Academy of Sciences the results of an extraordinary three-year program of research that had involved 236 balloon soundings to altitudes of between six and nine miles. His data proved beyond question that atmospheric temperature does not decrease with altitude indefinitely but levels off about seven miles above the surface. He assumed that the temperature remained constant throughout the atmosphere above that level. Without temperature differences, this upper region would be devoid of air movement, and Teisserenc de Bort concluded that, in the absence of wind and convection currents, the various component gases would settle into distinct layers, the heavier gases below the lighter.

Just as Robert Boyle had declared, 250 years earlier, that air is made up of several gases, Teisserenc de Bort now proposed that the atmosphere has distinct zones with different characteristics, and he proceeded to name them. The region below the seven-mile level, where temperature differences give rise to the winds and turbulence of the world's weather, he named the troposphere, using a Greek root meaning "to turn over." He called the upper atmosphere, where he expected strata of different gases to be found, the stratosphere. He was wrong about the dimensions and the nature of the stratosphere, but his labels endured.

Others rapidly elaborated on his revelation that the boundary of the troposphere, which came to be called the tropopause, is in effect a ceiling on the world's weather. As masses of air—called air parcels by atmospheric scientists—warm near the surface of the earth and rise through the troposphere by convection, they expand and grow colder because of the lower pressure aloft, just as air released from a pressurized can expands and chills.

In the troposphere, the temperature of the atmosphere surrounding the air parcels also decreases with height, simply because it is warmed from below by solar energy reradiated from the earth; as long as the rising air is warmer than the air around it, convection continues. But in the steady temperature of the tropopause, rising parcels are no longer warmer than their surroundings and rise no farther. As a result clouds, which develop vertically by convection, almost always are stopped at the tropopause; one manifestation of this collision is the flat anvil cloud atop a summer thunderstorm.

The existence of this thermal inversion, so called because it is an inversion of the normal decrease of temperature with height, is a consequence of the way the planet and its atmosphere absorb energy from the sun and radiate it back into space. The British astronomer Sir William Herschel had shown in 1800 that more is involved in this process than visible light. Herschel had passed a thermometer through the spectrum of colors produced by a prism and had seen the temperature rise as he moved from the violet to the yellow and on to the red. That was no surprise. But when he moved the thermometer beyond the red band, the temperature rose further. Invisible energy was being radiated there, and Herschel surmised that it was caused by what are now called infrared (beyond the red) rays.

That even more might be involved was suggested several decades later by one of the giants of 19th Century scientific thought, the Scottish mathematician and physicist James Clerk Maxwell. He showed that electricity is always accompanied by magnetism, and vice versa, and that an electrical current produces a corresponding magnetic field that oscillates like the current and extends outward from it. He concluded that the electrical and magnetic oscillations are in fact the same phenomenon, and called them, as a group, electromagnetic waves. Maxwell determined that these waves travel at approximately the same speed as light. Reasoning that the matching velocities could not be a coincidence, he decided that light must be an electromagnetic wave sent out by the sun. This in itself was a radical notion, but Maxwell went further.

Visible light is merely one part of a family of electromagnetic-wave forms, he said, that radiate from the sun at the speed of light. They differ only in their wavelength and frequency. Infrared light is an invisible radiation whose waves are slightly longer than those of visible light; on the other side of the visible spectrum is ultraviolet radiation, whose waves are shorter. Maxwell expected that both longer and shorter waves must exist.

In 1888, the German physicist Heinrich Rudolf Hertz confirmed and expanded on Maxwell's vision. Setting various kinds of electrical charges zapping through his laboratory, Hertz identified waves whose length was as much as a million times that of visible light. This not only fulfilled Maxwell's prediction of a broad electromagnetic spectrum but opened the way to a tremendous technological advance: Not long thereafter, Guglielmo Marconi in Italy found a way to use the Hertzian waves for wireless communication; they came to be called radio waves.

The discovery of the electromagnetic spectrum had profound implications for atmospheric study. Scientists had been given a unifying theory that explained all electromagnetic energy reaching the environs of the earth, from gamma rays with wavelengths measured in trillionths of an inch through light waves millionths of an inch long to radio waves more than five miles in length. With this knowledge scientists were able to quantify

Ever since the first manned balloon flight over Paris in 1783, lighter-than-air vehicles have aided in the study of the atmosphere. Many early balloon flights, like the sensational equestrian ascent pictured below, yielded little scientific information. Soon, however, aeronauts thought to carry aloft barometers, thermometers, compasses and other instruments. One researcher, French physicist Joseph Louis Gay-Lussac, soared to 23,000 feet in 1804 and retrieved an air sample that turned out to contain the same proportion of oxygen and nitrogen that air at the surface does.

Such high-level essays were not without risk. In 1862, English meteorologist James Glaisher lost consciousness in the thin air at 29,000 feet, surviving only because a frostbitten companion was able to open the balloon's gas valves and get them down. Yet the prospect of charting the weather at its source lured Glaisher back to the sky in a matter of months (*page 49*).

Sightings of shooting stars and other spectacular astronomical events demonstrated that balloons, ranging above dust, haze and smoke, could be first-class observatories. Even after interest in manned flights aboard the capricious craft waned early in the 20th Century, balloons remained eminently useful to science by lifting instruments to 75,000 feet or higher—permitting study of the previously inaccessible stratosphere.

Bundled against the chill 12,000 feet above Paris, Joseph Louis Gay-Lussac collects an air sample while colleague Jean-Baptiste Biot takes a temperature reading. Their pioneering research flight was made in August 1804.

A decoratively trimmed balloon lofts French aeronaut Pierre Léstu-Brissy and his mount in October 1798. One not-very-useful tidbit of knowledge resulted: The horse suffered nosebleeds at relatively low altitudes.

A balloon flown in April 1868 by two
French aeronauts, Camille Flammarion and
Eugene Godard, casts a brightly haloed
shadow at left in this contemporary illustration.
Flammarion marveled at the optical effect
(*page 34*) that reproduced the details of his
craft, even down to the "thinnest ropes."

A meteor shower graces the night flight
of a manned balloon over France in
November 1867. Rising above pollution
and clouds, balloons provided excellent
vantage points for astronomers.

A meticulously illustrated chart
presents the altitude and temperature
readings taken on July 21, 1863,
during meteorologist James Glaisher's
flight from the Crystal Palace in
London to Epping Forest, 20 miles away.

all the energy reaching the earth from the sun, track what happened to it and determine how it affected the atmosphere.

One of the consequences of their investigations was an explanation—offered in the 1930s—of the temperature inversion above the tropopause. Earlier, it had been calculated that roughly 50 per cent of all incoming solar radiation reaches the earth, where it is absorbed and radiated back into the lower atmosphere as heat. In 1930, the British geophysicist Sydney Chapman proposed that much of the ultraviolet radiation arriving from the sun is absorbed by the stratosphere, which gains considerable heat in the process. Between these two regions of heat absorption is the tropopause. Its altitude varies according to the amount of solar energy reaching and being radiated back by the earth: The tropopause is roughly 10 miles high at the Equator, where the earth radiates large amounts of heat; about seven miles high in temperate regions; and about five miles high at the Poles. Since the temperature of the troposphere decreases with altitude at about the same rate regardless of latitude, this variation in height causes a paradoxical result: The tropopause actually is colder at the Equator than at the Poles because it is farther from the earth's surface.

Although Teisserenc de Bort assumed that the stratosphere extended indefinitely into space and that its temperature was virtually constant, evidence soon emerged that it is in fact a relatively thin layer above which temperatures begin to change again. After World War I, an Oxford University physicist, Frederick Alexander Lindemann (later Winston Churchill's chief scientific adviser), and meteorologist G. M. B. Dobson became intrigued with meteors—the evanescent bits of interplanetary debris that streak through the nighttime sky. By applying the laws of thermodynamics to meteor trails, they hoped to determine atmospheric temperatures far above the 15-mile range of sounding balloons.

As a meteor—a bit of solid material consisting primarily of stone or iron—enters the atmosphere at speeds that may exceed 100,000 miles per hour, it is vaporized by frictional heat. Vapor molecules then strike the widely dispersed air molecules with such force that the air's constituent gases blaze in response (much as neon glows when excited by an electric current), painting a streak of light across the sky. Realizing that the altitude, length and brightness of a meteor's trail are all affected by the air's density, Lindemann and Dobson devised mathematical formulas to calculate the density, and from it the temperature, of the air through which a meteor had passed. They then applied the formulas to the assembled observations of a network of amateur astronomers, who simply judged by eye the brightness, length and height of hundreds of meteor trails.

In 1922, this technique, subjective and convoluted as it was, yielded a startling result—the temperature of the air about 30 miles above the earth seemed to be roughly 70° F. Previous researchers, assuming that temperatures decreased steadily up to the tropopause and remained constant in the stratosphere, had calculated that at 30 miles of altitude the reading would be below –60° F. If Lindemann and Dobson were correct, the stratosphere was being heated in an unknown, but significant, way.

Although their finding was greeted with considerable skepticism when it was published in the Royal Society's *Proceedings,* one British scientist, F. J. W. Whipple, immediately realized that it might explain a puzzling

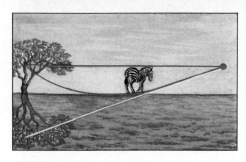

Reality and mirage coexist in this photograph of a zebra contemplating a mirage on a dusty Kenyan plain. In a diagramed side view of the scene, the camera *(red circle)* receives both straight and refracted light rays *(red lines)* from the distant trees. When the bent rays reach the lens, they are coming from the direction indicated by the yellow line, and they create the illusion of a reflected tree.

Mirages, such as the tantalizing but unattainable lake that appears in a burning desert, have been variously interpreted as supernatural events or the workings of a fevered mind. In fact, they are the straightforward products of the interaction of light and air.

A mirage occurs when light rays pass from one layer of air to another layer with a different density. In the process, their speed is altered and their course bent—or refracted—so that they strike the eye from a direction other than that of their point of origin. The difference in the air's density is caused by differing temperatures, and the light bends toward the colder, denser air.

If the air near the surface is much hotter than that just above it, as is often the case in a desert, light rays moving toward the ground from the top of a tree, for example, will bend up from the hot air to the cooler air at eye level. Light from halfway down the trunk will also bend upward, but at a smaller angle. The result will be a mirage in which the tree appears below its actual location and in an inverted position, like a reflection. At the same time, other light rays are traveling straight from the tree to the eye, so the tree will also be seen in its actual position.

Light from the sky may go through the same distortions, making a field of shimmering blue appear on the ground and creating the irresistible impression of trees reflected in a body of water. This kind of mirage can also appear on sun-baked highways.

When the air is cooler at the surface than at higher levels, as frequently happens over the cold waters of a lake, a mirage may appear above the real object. Light rays heading skyward refract down toward the colder air, with the result that a sailboat on the lake may seem to have a duplicate cutting a stately course across the sky.

Such distinct mirages occur only from time to time, but light rays are always being refracted by the ever-changing atmosphere. Mirages are thus a dramatic manifestation of the fact that almost nothing is exactly where it appears to be.

acoustical phenomenon observed during the Great War: The thunder of heavy guns sometimes was audible hundreds of miles from the front, yet the sound seemed to skip over an intervening "zone of silence."

Using reports in various magazines and journals, Whipple compiled maps of the audibility of major explosions—the firing of guns at naval reviews, detonations designed expressly for atmospheric research and various accidental explosions. Most of these maps showed two distinct zones of audibility separated by an area where nothing was heard; sounds reaching the outer zone, he found, did so between one and two minutes after the original explosion, but could not have gone in a straight line because nothing was heard in the middle zone. Whipple decided that the sound waves had somehow traveled upward from their source, over the zone of silence and back to earth in the outer zone. He was convinced that the temperature change described by Lindemann and Dobson was responsible.

Determining the exact path of the sound waves, however, required that the precise time of the initial explosion and distant observations be recorded. In the late 1920s, Whipple obtained this data by enlisting the cooperation of ordnance officers from the Royal Arsenal at Woolwich, where heavy guns routinely were tested, and of the British Broadcasting Corporation. During selected firings, the Woolwich range officer transmitted a signal over the BBC radio network as he triggered each blast. Six stations, situated in a half circle anywhere from 105 to 165 miles from Woolwich, used special apparatus to record both the BBC signal and the sound waves bouncing down from the stratosphere. The results confirmed Whipple's beliefs. Although the sound waves from 15- and 16-inch naval guns were inaudible to humans at great distances, they clearly registered on his sensitive microphones; even firings of the 8-inch guns aboard light cruisers could be detected. Whipple was able to calculate that the sound waves reached altitudes of 25 to 37 miles—roughly the altitude of Lindemann and Dobson's warm layer—before bending back to earth.

Whipple knew that sound travels faster in warm air than in cold and that the leading edge, or wave front, of a sound moving outward from its source often encounters different temperatures as it rises. Usually, the higher air is colder and slows the upper edge of the wave front, thus bending the entire sound upward. But when a wave front meets the warmer air 30 miles up, he reasoned, its upper edge outdistances the lower, gradually bending the sound wave back to the earth. The zone of silence exists between the point where all the sound waves have been refracted upward away from the surface and the point where the first waves return from the stratosphere.

The behavior of the sound waves was understood before anyone knew what caused the warm layer in the upper stratosphere. In the 1930s, an answer to that mystery was found. Researchers related the heating to ozone, a form of oxygen. Ozone is created when short-wave ultraviolet radiation splits an ordinary two-atom oxygen molecule; the chemically active single atoms attach themselves to ordinary oxygen molecules to form three-atom molecules of ozone. This process generates an unusually high concentration of ozone in a layer that completely surrounds the earth between altitudes of about six and 30 miles. The gas is explosive and quite toxic in high concentrations, but poses little threat to humans; ozone is virtually absent in the troposphere because it decomposes when it touches a solid, such as the surface of the earth. Concentrations are highest at altitudes of between 12 and 18 miles, but even there it constitutes only .00001 per cent of the atmosphere—barely enough to form a layer of pure ozone $1/10$ inch thick at normal sea-level pressure and temperature.

Despite its sparsity, ozone profoundly affects atmospheric temperature—and life on earth—because it absorbs a potent segment of the solar spectrum, ultraviolet radiation. The relationship between the gas and this particular emission is complex. Having been created by ultraviolet radiation, ozone then absorbs the rays; the absorption can destroy the ozone molecule by splitting off its extra oxygen atom. The creative and destructive effects are roughly in balance at the top of the ozone layer. At lower levels, more of the rays are filtered out by the ozone above and less of the gas is destroyed: Thus the stratosphere is warmest (perhaps 65° F.) near the top of the ozone layer, where the intense energy of the ultraviolet rays is expended, rather than at its center, where the ozone concentration is greatest. Very little of the ultraviolet radiation makes it to the earth's surface, and this filtering effect is fortunate indeed; without it the earth would probably be uninhabitable: Plants would wither; animal life on the surface, including man, would suffer burns, skin cancer and blindness; and the oceans would warm until marine life as it now exists would be untenable.

At the ozone layer's hot upper edge, about 30 miles above the earth, lies the stratopause, the boundary between the stratosphere and yet another region—the mesosphere. Although the mesosphere can be explored by rocket, direct measurements from a fast-moving projectile are exceedingly difficult to accomplish. One solution is somewhat reminiscent of the early Lindemann-Dobson meteor studies; rockets ferry small spheres into the upper atmosphere, then drop them. Scientists track the falling bodies with radar and record speed and acceleration data; this enables them to determine the air resistance a sphere encounters, and thus calculate air density and temperature. Another method employs rocketborne grenades detonated in the upper atmosphere; the velocity of the explosion's sound waves can be used to calculate the air density, temperature and wind speed.

Such experiments, conducted in the late 1940s and early 1950s, showed that the rarified gases above the stratopause absorb so little solar heat that the 20-mile-thick mesosphere rapidly gets colder with increasing height. That much was to be expected. But the experiments also showed that, paradoxically, the region is coldest during the summer months, apparently because of some mysterious global exchange of warm and cold air.

When the mesosphere is at its coldest, clouds unlike anything in the lower atmosphere may be formed in high-latitude regions. Called noctilucent clouds because they are illuminated by the sun yet are visible only in the dark sky after sunset, these silver white veils gather about 50 miles above the earth at the mesopause, the coldest region in the entire atmosphere, where summer temperatures sometimes drop to −225° F. Rockets have retrieved samples of these clouds, but their composition remains uncertain: Some researchers believe that the clouds are made of microscopic dust from meteors, comets or volcanoes; others maintain that dust merely provides nuclei for ice crystals (a difficulty with this hypothesis is that the atmosphere at such heights apparently contains little water vapor).

The density and temperature of the air in the layer of the atmosphere that extends from the 50-mile-high mesopause to perhaps 180 miles above the earth—the thermosphere—were not directly measured until the late

1950s, when scientists began to calculate changes in the orbits of satellites caused by frictional drag. Such studies revealed that although the thermosphere contains very little air—less than $\frac{1}{100,000}$ of the atmosphere—what air there is absorbs most of the incoming solar radiation in the extremely short ultraviolet range. The result is wild variation between daytime and nighttime temperatures. Under certain circumstances the temperature in the upper thermosphere may reach 3,600° F. just past midday, only to drop 1,000° F. or more at night as the thin air rapidly loses heat. Such extremes have little practical significance, however. At high altitudes the temperature—a measurement of the average speed of randomly moving gas molecules—becomes an abstract concept, shorn of its everyday meaning, because the air is so thin that it transfers virtually no heat to satellites.

Most of the thermosphere's heat is generated when solar ultraviolet radiation breaks molecules down into their constituent atoms. While the lower regions of the atmosphere are composed mostly of molecular oxygen and nitrogen, in the thermosphere this photochemical breakdown creates a complex potpourri of oxygen atoms, nitrogen oxides and nitrogen molecules, and a few hydrogen atoms. In the thermosphere's calm upper reaches, where neither winds nor convection mixes the various elements, these components stratify—just as Teisserenc de Bort had supposed: The heavier ones, such as molecular nitrogen and oxygen, settle to form layers below those of the lighter atomic gases, such as helium and hydrogen.

In addition to its chemical and thermal layers, the atmosphere possesses electromagnetic layers—a possibility barely contemplated before the 20th Century. Sailors had noted that compass needles sometimes strayed unaccountably, and a few theoreticians had suggested that the sun's heat might somehow make the entire atmosphere behave like a huge dynamo. But not until the advent of radio did the existence of the atmosphere's electromagnetic layers become known.

In 1901, Guglielmo Marconi, the self-taught young Italian who already

A rare noctilucent cloud appears over the horizon near Plymouth, England. These clouds, which form five times higher than any other cloud, are so named because they can be seen only when lighted by the sun when it is over the horizon—that is, at night. The bright spots beneath this cloud are surface lights.

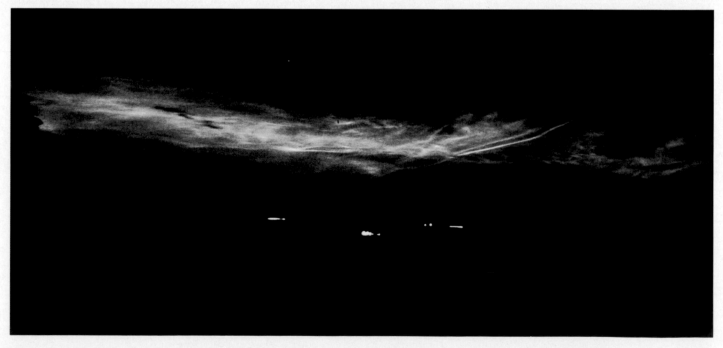

had built the first workable wireless, started to experiment with long-distance radio communication. The same year, he succeeded in sending a Morse code transmission from Cornwall, England, across the Atlantic Ocean to Newfoundland. How the radio waves had followed the curvature of the earth over such a distance eluded a number of electrical engineers and physicists who tried to explain Marconi's feat.

Several months later, a reclusive English physicist named Oliver Heaviside offered a hesitant, much-qualified explanation that proved to be correct: "There may possibly be a sufficiently conducting layer in the upper air. If so, the waves will, so to speak, catch on to it more or less. Then the guidance will be by the sea on one side and the upper layer on the other." Heaviside gradually buttressed his theory with computations; meanwhile, a British-American electrical engineer, Arthur Edwin Kennelly, a former assistant to Thomas Edison, independently arrived at the same conclusions.

Today the conducting layer is called the ionosphere, and is known to permeate the upper mesosphere and the thermosphere. As X-rays and ultraviolet radiation from the sun whip into these regions, they strip electrons from atoms and molecules. The electrons carry a negative electric charge and are caught up in the oscillating electromagnetic field of any passing radio wave. When moving, a charged particle emits its own electromagnetic energy; in effect, each electron becomes a tiny transmitter, whose signal has the effect of bending the original radio wave. The more free electrons present, the more the radio wave is affected, and enough of them will bend the radio-wave front earthward in the same manner that a sound wave is refracted by a warm air layer aloft.

There is a direct relationship between the number of electrons present in the air and the electromagnetic frequency they affect: The more densely packed the electrons, the higher the frequency they refract. Since ionization occurs primarily because of incoming solar radiation, the number of electrons increases with altitude—to a limit of about 185 miles, where electron density is highest and the highest frequencies are refracted. It is not an even progression, but occurs in steps, and scientists have labeled a number of regions in the ionosphere where particular frequencies are most affected.

The extent of the regions can only be approximated, since some of them weaken or disappear at night, or in winter, but scientists generally recognize three: the E region, between 55 and 100 miles above the earth; the D region, below 55 miles; and the F region, above 100 miles. The F region is often subdivided into F_1 and F_2 regions, because the uppermost region—the F_2, beginning about 155 miles up—normally has the highest concentration of electrons. Even here, however, the density of electrons is never great enough to affect frequencies above 15 million cycles per second.

The D region contains relatively few free electrons but, owing to the effect it has on certain radio waves, is of special interest. Normally radio transmissions in the AM band of frequencies—from 550,000 to 1.6 million cycles per second—are not refracted until they reach the E region. But during the daytime they are absorbed in the D region, where the air itself is more dense and free electrons collide more frequently with heavier atoms and molecules than they do in the thinner air above. When such collisions occur, the electromagnetic energy of the moving electron is lost, not only to itself but to the radio wave, which weakens or disappears. At night, however, when there are far fewer electrons—and collisions—in the D region

Although invisible in this photograph of the orbiting *Apollo 9* command module, the upper reaches of the atmosphere pose a tangible threat to the capsule's safe return to earth.

Piercing the Earth's Ethereal Armor

Space vehicles streaking toward earth at speeds of roughly 25,000 miles per hour are, in effect, man-made meteors. And just as the atmosphere shields the planet from space debris, so too does it present a truly formidable barrier to homeward-bound spacecraft; they must enter the atmosphere at a shallow angle to ensure that the air will slow them without causing a disastrous build-up of frictional heat.

The optimal angle for reentry—first calculated for the manned space flights of the late 1960s—is approximately 6 degrees, relative to the edge of the thermosphere, about 75 miles above the earth. At a smaller angle of approach, the thin air would create enough lift on the lower surface of the spacecraft to loft it back into deep space. With an approach angle any steeper, the intense friction caused by the capsule's high-speed collision with air molecules would create enough heat to melt heat shields —and even the spacecraft itself—in a matter of minutes.

The *Apollo 13* lunar-landing and service modules, jettisoned by the returning astronauts after their 1970 moon mission was aborted, are incinerated by friction-caused heat as they plummet through the atmosphere.

In this schematic diagram, a space vehicle that is attempting to reenter the atmosphere at too shallow an angle is deflected back into space, and one descending at too steep an angle is burned up by friction. For a safe return, a spacecraft has to be within 2 degrees of the optimal, 6-degree reentry angle.

radio waves pass through it unaffected and are reflected from higher up, making reception of distant AM stations possible.

Television and FM radio waves are of too high a frequency to be reflected by the ionosphere and pass directly out toward space; TV transmission over great distances must therefore be assisted by cable, microwave relay or satellite relay, although freak bounces off the ionosphere occur occasionally. The effects of radio-wave refraction depend on the angle at which the signal enters the ionosphere. As the path of the radio wave approaches the perpendicular, it reaches a point where it cannot be refracted back to earth and it continues into space. Thus the refracted signal of a radio station 200 miles away may be heard clearly while one 60 miles off cannot be picked up. Refraction is also affected by disturbances in the ionosphere. Because solar radiation is the key to ionization, solar disturbances such as sunspots and solar flares play havoc with radio reception. At such times the aurora, whose unique displays are staged in the ionosphere, lights up with special brilliance, and ham-radio operators find their receivers crackling with static. During solar flares the lowest ionized layer temporarily becomes very absorptive, and long-distance radio reception may be interrupted for hours.

Aside from its effects on communication, the ionosphere makes its presence felt in another way: It glows. The airglow is caused by the same electrochemical reaction that causes ionization; instead of merely stripping an electron from a molecule, short-wave radiation from the sun may change the energy state of the molecule, causing it to release some of its energy as light. The sky's luminosity on clear, moonless nights is due primarily to airglow, which is brighter than the light of all the stars put together.

In 1956, physicist Murray Zelikoff found in laboratory experiments that when he introduced nitric oxide into a sphere containing the same mixture of gases that exists above the stratosphere, a chemical reaction took place and released extra radiation. Perhaps, he thought, the nitric oxide present in the upper atmosphere provides one of the sources of airglow, and if so, then increasing the concentration of the gas would heighten the airglow. On a clear, starlit night, Zelikoff fired a rocket carrying glass containers of nitric oxide over the wilds of New Mexico. Sixty miles up, the rocket released its 20-pound payload of gas. Moments later his colleagues, peering through telescopes atop a hill about 60 miles distant, saw a round, yellow glow in the sky. It grew to four times the size of a full moon and gradually changed color from gold to silver gray, glowing for fully 20 minutes.

Zelikoff's success suggested to some that artificial airglows could be used to light cities at night, speed the growing of crops or illuminate nighttime rescue missions. Fortunately the possibilities were not pursued. The nitrogen oxides, drifting downward, could wipe out the ozone layer, inflicting incalculable damage on the earth.

The ionosphere extends almost to the top of the thermosphere, where the air molecules become so sparse as to be negligible. Above it is the farthest province of the atmosphere—the exosphere, a bleak environment irradiated by a pitiless sun. While air molecules at sea level can travel only about $3/1,000,000$ inch before colliding with another, the light gases of the exosphere—helium, atomic hydrogen and atomic oxygen—are so widely dispersed that an atom travels for about six miles before colliding with another. With fewer collisions to slow them, the atoms, heated to perhaps 3,600° F., move several times faster than molecules in the lower atmo-

A diagram shows how the lowest level of the ionosphere, the D region (*blue*), affects AM radio transmissions. During the daytime, solar radiation turns the layer into a seething mass of ions and freed electrons that collide with incoming radio waves and absorb them.

At night, D-region activity subsides. AM radio waves penetrate to the E region (*black*), which reflects some of them back to earth, where they may be detected at great distances. Some AM radio stations have to end their transmissions at dusk to avoid interfering with remote stations on the same frequency.

sphere, spraying up toward space like water from a fountain. Gravity eventually pulls most of the speeding atoms back toward earth, but a minute fraction of the helium and hydrogen atoms, moving at speeds in excess of 25,000 miles per hour, escape into interplanetary space. There is no danger of the atmosphere leaking away, however: Lost helium is replenished by the radioactive decay of rock on earth, and hydrogen generated by the photochemical breakdown of water vapor (at altitudes of about 65 miles, where it tends to become stratified) more than compensates for the losses.

Just where the exosphere ends and interplanetary space begins cannot be defined with any certainty, for it is merely a question of degree; molecules become more and more scarce until, somewhere between 300 and 900 miles above the earth's surface, there are none.

Until the late 1950s, it was generally assumed that beyond the exosphere lay a void—a vast realm of nothingness, pierced only occasionally and momentarily by showers of charged particles from the sun, cosmic rays from interstellar space and plummeting meteors. This description, inferred by scientists who had been unable to obtain direct measurements of conditions at altitudes of more than about 120 miles, was utterly demolished by the flight of the first United States satellite, *Explorer I,* in 1958.

Although smaller than the Sputnik satellites previously orbited by the Soviet Union, *Explorer I* was packed with sophisticated instruments assembled largely under the direction of James A. Van Allen, a physicist from the State University of Iowa who had spent a decade studying radiation and magnetic fields in the atmosphere. Along with other scientists, he had begun to suspect that space was not empty—that it was, in fact, alive with a radiation produced not by waves of energy but by charged particles. From an orbit that reached as high as 1,500 miles, the 31-pound satellite presented what Van Allen described as "a broad, deep stripe of knowledge 4,700 miles wide and 1,400 miles thick around the waist of the world."

At first, *Explorer I's* radiation counter, which measured the frequency of impacts by charged particles, recorded the kind of pattern that Van Allen and his team expected. During its orbits over the United States, as the satellite probed some 250 miles into space, it detected more and more charged-particle radiation. Later, however, at altitudes high above the Equator, the readings suddenly dropped, often to zero. The dismayed scientists assumed that the radiation counter was malfunctioning.

A second Explorer satellite presented the same anomaly. While puzzling over its data, one of the scientists, as Van Allen remembered, "called our attention to something we all knew but had temporarily forgotten." Extremely high radiation could jam the instruments and produce a reading of zero. One of the team exclaimed, "Space is radioactive!"

That was, of course, an overstatement, but the truth that emerged from the continuing exploration of space was hardly less intriguing. What had been assumed to be a void proved to be comparatively teeming with matter; and what had been thought of as the end of the earth's protective envelope—the exosphere—was found to be surrounded by a far larger aura that is not, strictly speaking, part of the atmosphere but is inextricably bound up with its workings and with life on earth.

As Van Allen and others had suspected, a wind of sorts is blowing through space—a tenuous but constant stream of tiny particles flashing outward from the sun at speeds averaging one million miles per hour. This

Unmasking the Aurora

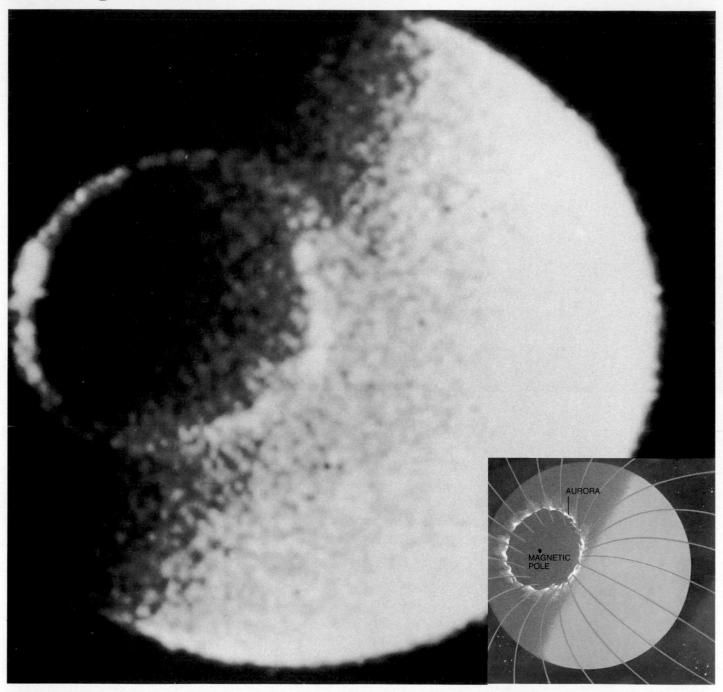

AURORA

MAGNETIC POLE

Until recently, any view of the auroras that flicker in the polar skies, however awe-inspiring, has been fragmentary. Sightings of the glowing draperies and other typical auroral shapes inevitably are skewed by the viewer's earthbound perspective, 50 miles or more below the aurora, or are cut off at the horizon.

Only from space can the aurora be seen for what it is. In the vicinity of each Pole, the funnel-like alignment of the earth's magnetic lines of force acts to guide charged particles of the solar wind toward the surface of the earth until they collide with the molecules of the atmosphere. The result is a great halo of glowing air particles centered on the magnetic pole.

The shape of an entire aurora—the ring of light visible on the earth's night side—was captured for the first time in this photograph, taken from a satellite on September 15, 1981. The inset drawing depicts the magnetic lines of force that give the aurora its form.

solar wind—consisting mostly of positively charged protons and negatively charged electrons of hydrogen and helium, collectively called plasma—is a light breeze indeed; in deep space, it contains about 90 particles per cubic inch, whereas at sea level on earth there are 443 quintillion (that is 443 followed by 18 zeros) atmospheric particles per cubic inch. Still, the solar wind exerts real force. It accounts for the fact that the dusty tail of a comet is often seen to extend not backward along the comet's line of travel, but in a direction away from the sun, blown askew by the solar wind. The existence of the solar wind has prompted hopes that one day man can sail through space, using gigantic sheets of foil to run before it.

Once the presence of a continuous solar wind had been established beyond dispute, scientists confronted the corollary question: What prevents it from irradiating the earth? Gusts of the solar wind sweep into the immediate environs of earth only infrequently, and the resulting disruption of radio communications and magnetic compasses, along with the dazzling eruptions of the northern and southern auroras, testifies to the devastation that would be wrought should the solar wind not be somehow deflected.

The shield against that devastation, as Van Allen discovered, is provided by the earth's magnetic field—the invisible lines of force that, close to the earth, loop from one magnetic pole to the other and, farther out in space, are distorted by the solar wind into a kind of elongated teardrop. Initially, scientists thought that the magnetic field was most effective in two narrow belts centered above the Equator. One of the bands (both of which came to be known as Van Allen belts and, later, Van Allen zones) is between 600 and 3,000 miles up; the second and larger is found from 6,000 to 40,000 miles up. When, through complex processes that scientists do not fully understand, some of the high-energy particles of the solar wind enter these belts, their charges interact with the magnetic lines of force, and they become trapped; they spiral along the lines of force, from one pole to the other, ricocheting back and forth, as one writer has put it, "like prisoners condemned forever to run up and down spiral staircases."

Beyond the edges of the Van Allen belts, where the magnetic lines of force descend toward the magnetic poles, the charged solar-wind particles may smash into the ionosphere, lighting the sky with the aurora. As scientists gathered more data from farther out in space, they found that the plasma particles trapped in the Van Allen belts, combined with those that excite the aurora, amount to only a tiny fraction of the particles that stream earthward from the sun; the rest are deflected at the edges of the earth's magnetic field, or magnetosphere.

At the boundary between the magnetosphere and the solar wind, the magnetopause, the pressure between the plasma particles in the magnetosphere and those in the solar wind are balanced. The border is not rigid; it breathes in and out in response to changes in the force of the solar wind. The earth whirls through space within "an elongated cavity hollowed out of the solar wind," as Van Allen put it, which streams out from the sun like water and flows around the magnetosphere as if enveloping a broad-beamed torpedo. More than 40,000 miles sunward from the earth is a sort of bow wave where the impact of the plasma presses the magnetopause toward the earth. On the side of the planet opposite the sun, the magnetosphere trails away in a long taper until the flow of the solar wind rejoins, somewhere in space far beyond the orbit of the moon. **Ω**

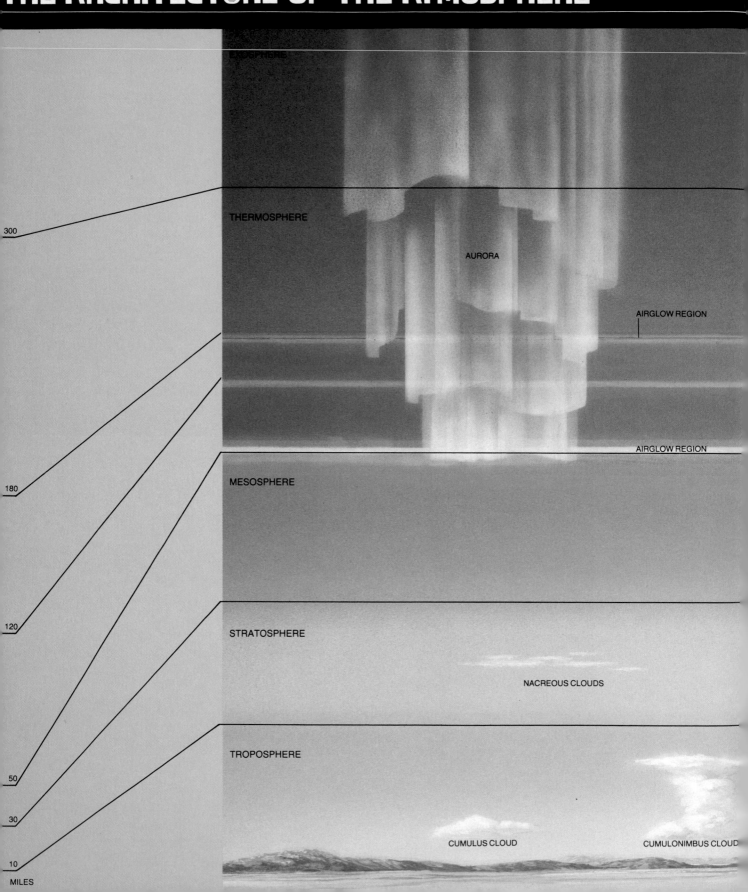

EXOSPHERE

THERMOSPHERE

AURORA

AIRGLOW REGION

AIRGLOW REGION

MESOSPHERE

STRATOSPHERE

NACREOUS CLOUDS

TROPOSPHERE

CUMULUS CLOUD

CUMULONIMBUS CLOUD

300

180

120

50

30

10

MILES

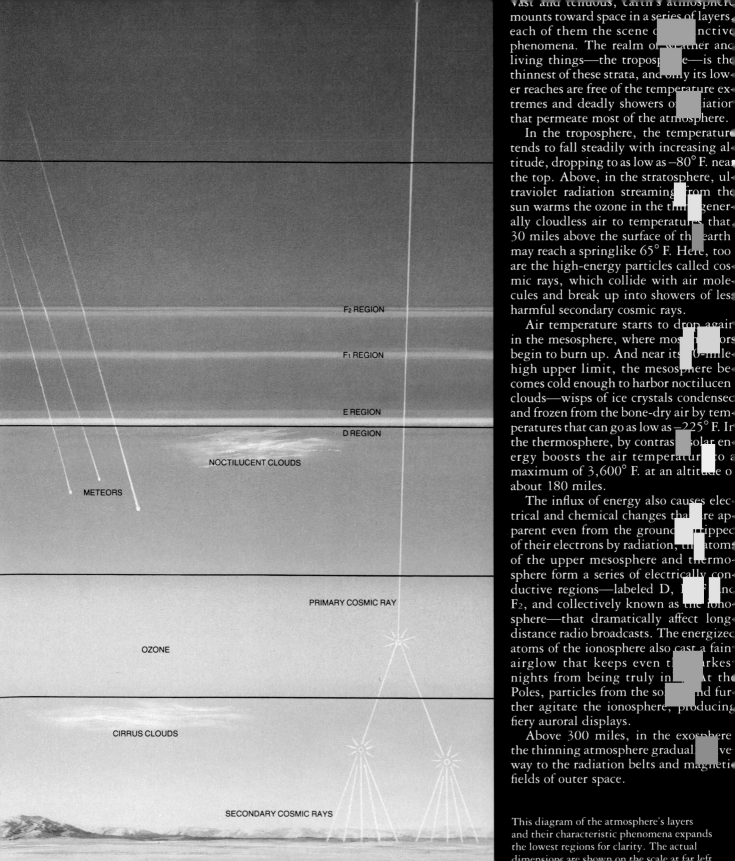

F₂ REGION

F₁ REGION

E REGION

D REGION

NOCTILUCENT CLOUDS

METEORS

PRIMARY COSMIC RAY

OZONE

CIRRUS CLOUDS

SECONDARY COSMIC RAYS

Vast and tenuous, earth's atmosphere mounts toward space in a series of layers, each of them the scene of distinctive phenomena. The realm of weather and living things—the troposphere—is the thinnest of these strata, and only its lower reaches are free of the temperature extremes and deadly showers of radiation that permeate most of the atmosphere.

In the troposphere, the temperature tends to fall steadily with increasing altitude, dropping to as low as −80° F. near the top. Above, in the stratosphere, ultraviolet radiation streaming from the sun warms the ozone in the thin, generally cloudless air to temperatures that 30 miles above the surface of the earth may reach a springlike 65° F. Here, too, are the high-energy particles called cosmic rays, which collide with air molecules and break up into showers of less harmful secondary cosmic rays.

Air temperature starts to drop again in the mesosphere, where most meteors begin to burn up. And near its 50-mile-high upper limit, the mesosphere becomes cold enough to harbor noctilucent clouds—wisps of ice crystals condensed and frozen from the bone-dry air by temperatures that can go as low as −225° F. In the thermosphere, by contrast, solar energy boosts the air temperature to a maximum of 3,600° F. at an altitude of about 180 miles.

The influx of energy also causes electrical and chemical changes that are apparent even from the ground. Stripped of their electrons by radiation, the atoms of the upper mesosphere and thermosphere form a series of electrically conductive regions—labeled D, E, F₁ and F₂, and collectively known as the ionosphere—that dramatically affect long-distance radio broadcasts. The energized atoms of the ionosphere also cast a faint airglow that keeps even the darkest nights from being truly inky. At the Poles, particles from the sun and further agitate the ionosphere, producing fiery auroral displays.

Above 300 miles, in the exosphere, the thinning atmosphere gradually gives way to the radiation belts and magnetic fields of outer space.

This diagram of the atmosphere's layers and their characteristic phenomena expands the lowest regions for clarity. The actual dimensions are shown on the scale at far left.

A Magnetic Shelter from the Solar Wind

In addition to its gift of light and warmth, the sun emits deadly radiation and a solar wind of protons and electrons that streams through space at one million miles an hour. X-rays and most ultraviolet rays—both inimical to life—are absorbed by the earth's atmosphere. The solar wind is deflected in deep space by another kind of mantle.

Because the solar-wind particles are charged, they cannot cross the lines of magnetic force that loop far out into space from the earth's magnetic poles. The torrent of particles encounters the magnetic field about 40,000 miles from the earth, parts and flows around it. The magnetosphere—the sheltered region that results—is distorted by the solar wind into the form of a comet, with a tail extending some 240,000 miles downwind of the earth. Some of the solar wind leaks into the funnel-shaped openings known as the polar cusps. Channeled toward earth, such particles create the magnificent auroras or enter the regions of trapped particles known as the Van Allen radiation zones.

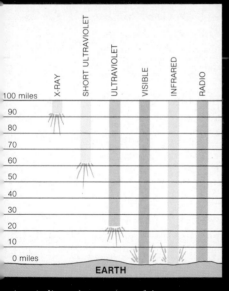

The chart indicates how regions of the upper atmosphere filter harmful X-rays and ultraviolet light from the sun's radiation but let visible, infrared, and radio emissions through. Except for the low-energy radio waves, the radiation heats the protective regions as it is absorbed.

A magnetic envelope, the magnetosphere, shelters the earth from the ionized blast of the solar wind. The impact on the magnetopause, the boundary of the magnetosphere facing the sun, creates a shock wave and a region of compressed and heated solar wind, the magnetosheath. The magnetosphere's long tail, which extends downwind, is not shown.

MAGNETOSPHERE

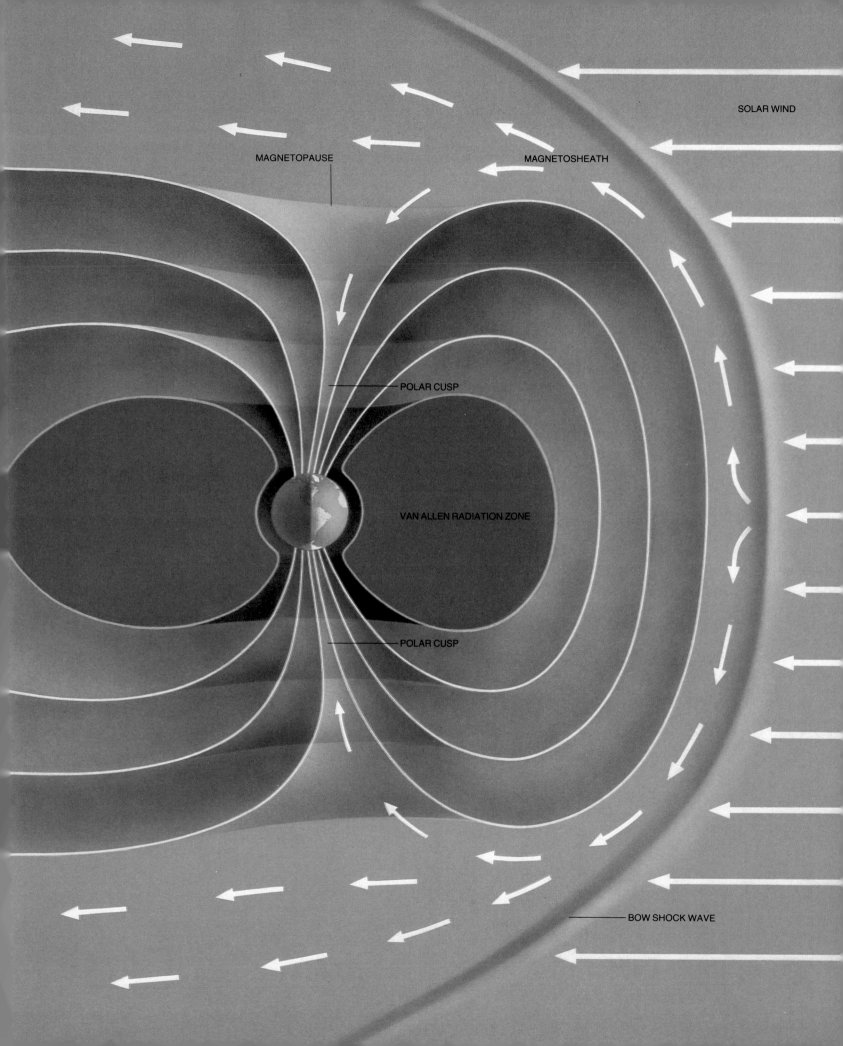

SOLAR WIND

MAGNETOPAUSE

MAGNETOSHEATH

POLAR CUSP

VAN ALLEN RADIATION ZONE

POLAR CUSP

BOW SHOCK WAVE

Chapter 3

THE MYSTERIES OF WIND

After the first gust, the wind kept up a continuous, monotonous scream, like the sound of escaping steam from a million engines, dreadful to hear, and the hut shook violently." So said the Russian writer I. W. Shklovsky in an account of his visit to the Yakut tribe in northeastern Siberia in the 19th Century. Although the Arctic wind was blasting the moss filler out of the cracks in the walls of the log cabin, Shklovsky wanted to see for himself what conditions were like out in the open. His hosts, who venerated this wind they called "the Chief," strongly urged him to stay put. "Don't go out," they warned, "the Chief will not like it."

But at Shklovsky's insistence his companions fastened a long strap around his waist and held the end tightly as he stepped outside. "Hardly had I opened the door," he wrote, "when I was flung violently to the ground. The air was full of ice crystals, which whipped my face like molten metal, and in an instant my cheekbones were frozen. I was as one blind in the impenetrable darkness. Where could I go? Where was the door? Had it not been for the strap held so firmly by my friends, it would have gone hard with me, for I could not possibly have found the door; but I managed to crawl on all fours to the threshold." As the stunned Shklovsky crept back into the hut, his hosts eyed him with amusement. "Well, friend," they asked, "what did the Chief say to you?"

The brutal lesson offered that day has been repeated continually for centuries throughout the world: Never underestimate the wind. Like the Chief, many recurring regional winds are so powerful and have such an impact on human life that they have been given special names, and stories of their power have been passed along from generation to generation. In Arabia, the searing *simoom,* or "poison wind," roars across the parched desert. In Turkestan, the *tebbad*—"fever wind"—scours the wilds. From the Sahara, the blistering sirocco brings misery to millions of Africans and blows dust and grit all the way to Europe. In France, the mistral delivers a penetrating, dry cold to the Rhone Valley every year. Although at times harmless and even refreshing, California's hot Santa Ana wind desiccates so much vegetation that it sets the stage for frequent, disastrous brush fires. Notorious all, these winds derive their unique identities from unusual local topography and atmospheric conditions, yet they are part of a magnificent global flow that is on the whole beneficial, indeed essential, to life on earth.

The wind is a great life-giver; it distributes the rains, moderates temperatures, cleanses the atmosphere and even broadcasts the seeds of many of the world's plants. Asia's great monsoons bring the rain required by crops

As Hurricane David passes 50 miles offshore, its peripheral winds lash pedestrians and palm trees on Miami Beach in 1979. David's winds reached a ferocious 150 miles per hour—yet the hurricane represented only about $\frac{1}{10,000}$ of the kinetic energy in the atmosphere.

that feed nearly half of the world's population, and for centuries the Atlantic trade winds made possible the flourishing commerce between Europe and the New World. Loved or loathed, the wind is everywhere. "There are some whole countries where it never raines, or at least very seldome," wrote the English philosopher Sir Francis Bacon in 1622, "but there is no Countrey where the Winde doth not blow, and that frequently."

Early attempts to understand the source of this universal experience relied on such mythical causes as the flapping of angels' wings or drafts from the cave of the Greek god Aeolus. Later attempts to apply reason alone left much to be desired; even the great Greek logician Aristotle could only postulate that the winds were essentially the dry sighs of a breathing earth. "It is absurd to suppose," he declared flatly, "that the air which surrounds us becomes wind simply by being in motion." In fact, wind is precisely that—air in motion.

Knowledge about global wind systems did not emerge in an orderly or systematic way; one early, and typical, fragment was the result of a voyage of discovery that began at dawn on August 3, 1492, when three small ships commanded by Christopher Columbus weighed anchor and set sail for the Orient. Columbus reckoned that his goal lay some 2,400 miles to the west, but instead of sailing due west from Spain, he shaped a course to the south. This first 800-mile leg of his voyage took him down to the Canary Islands, off the coast of Africa. Here, he knew, Portuguese ships coasted the continent on steady northeasterly winds, and it was his fervent hope and shrewd

Apple-cheeked cherubs puffing sea breezes adorn a map published in 1544. It traces an early voyage to Peru and the globe-girdling route of Ferdinand Magellan, who relied on a more realistic knowledge of wind power than the mapmaker displayed.

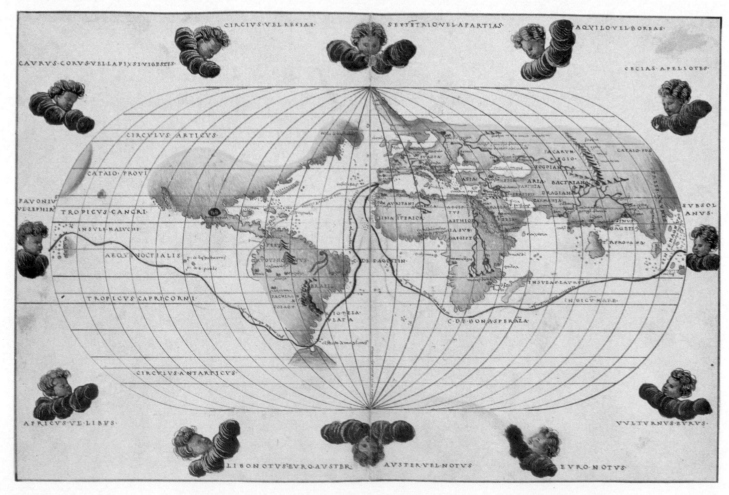

68

guess that these winds would propel his little fleet westward across the great, unexplored ocean. He had guessed right, and 36 days out of the Canaries he made landfall in the New World.

For nearly three months Columbus searched the Caribbean for the court of China's Great Khan, never suspecting that the Khan lived some 9,000 miles farther west. Finally, he gave up and headed for home. Knowing that he could not buck the easterlies that had brought him to the new land, he sailed northward, hoping to find a westerly wind to blow him back to Europe. Again his intuition paid off. After a two-week sail to the northeast, he found his westerlies and rode them all the way to Portugal. Columbus did not realize it at the time, but in addition to new lands, he had discovered one of the earth's great wind systems. The reliable north-easterlies that girdle the Northern Hemisphere near the Equator came to be known as the trade winds, for an archaic word meaning "course" or "track." Below the Equator, a mirror-image wind system creates similar southeasterly trade winds.

Mariners made immediate and frequent use of the trade winds, but nearly 200 years passed before anyone offered a reasonable explanation for their existence. Once again, the advance—achieved by the eminent English astronomer Edmond Halley—arose from an investigation of an entirely different subject. While sojourning in the South Atlantic to catalogue the stars of the Southern Hemisphere, Halley had ample opportunity to observe and become intrigued by the behavior of the trades. In 1686, nine years before he began calculating the orbit of the comet that bears his name, Halley set down a theory about these fascinating winds. Near the Equator, he said, where the sun's heat is more direct and intense than elsewhere on the globe, air rises "towards those parts where it is more rarified," and the trade winds flow in to replace it. Although this much of Halley's theory was sound, he went astray when he tried to explain why the trade winds flow from east to west. He decided that the rising air followed the westward movement of the sun and that the cooler air swirled in behind it in the form of easterly winds.

The real reason for the east-to-west flow of the trade winds continued to elude European scientists until a Londoner named George Hadley deftly modified Halley's model. Although a lawyer by profession, Hadley, like many other educated men of his day, devoted much of his spare time to the natural sciences. In 1735, he presented a paper on the trade winds to the Royal Society of London, one of Europe's august scientific groups.

Hadley agreed with Halley that the heated air rises at the Equator and then moves out toward the Poles, cooling gradually in the upper air. As it cools, he said, the air sinks and is drawn back to the Equator by the equatorial updraft. This cell-like circulation, he said, exists in both hemispheres. But instead of being drawn westward by the sun's heat, according to Hadley, the wind only appeared to come from the east. It was well-known that the earth rotates toward the east at high speed, and Hadley reasoned that the atmosphere must rotate along with it; he suggested that, because the surface of the planet rotates with the greatest speed where its circumference is greatest, at the Equator, air arriving from the Poles would lag behind the earth's surface, and would thus appear to be arriving from an easterly direction.

A carrousel provides an analogy for Hadley's concept. If a person stand-

A billowing wall of dust several thousand
feet high sweeps across a Syrian desert. Such a
squall, known as a haboob, is spawned when
a monsoon collides with dry air currents above it.

ing on the hub of a merry-go-round tosses a ball to a companion riding on the rim, the ball travels in a straight line, as any observer not on the carrousel could readily confirm. But the person on the merry-go-round trying to catch the ball would see it curve dramatically away, and would have to stretch to catch it. For the same reason, anyone standing on the whirling earth perceives the wind as curving; north of the Equator the curve results in a wind from the northeast, and south of the Equator, from the southeast—precisely the pattern of the trade winds.

Hadley also accounted for the westerly winds outside the tropics. If the air at the Equator was moving from west to east faster than the air near the Poles, he reasoned, then the air flowing toward the Poles would be moving eastward faster than the surface of the earth itself. Therefore the wind in the middle latitudes would appear to blow from the west. The explanation was a useful one, and more than a century passed before scientists realized that the concept of a single convection cell extending from the Equator to the Poles was short of the mark.

What led to the eventual solution was a quiet achievement in mathematics. In 1835, a brilliant French scientist named Gaspard Gustave de Coriolis published a theory describing in detail the behavior of bodies in motion on a spinning surface. By means of unshakable mathematical calculations, Coriolis proved that the path of any object set in motion on a rotating body curves in relation to any other object on the rotating surface. This is true, he showed, no matter where on the surface or in what direction the object is set in motion.

One of the first to see the connection between Coriolis' esoteric mathematical treatise and Hadley's notions on wind deflection was William Ferrel, an American schoolteacher and a dedicated student of the winds. In 1855, Ferrel became acquainted with a tremendous store of information on global winds and ocean currents, data that had been carefully compiled by the United States Navy's chief hydrographer, Lieutenant Matthew F. Maury. Amassing reports from ships' captains, the Naval officer had discerned hitherto-unsuspected wind-circulation patterns and areas of relatively constant barometric pressure. Most notably, Maury found that there are belts of constant low-pressure air near the Equator and the Poles, and that a zone of stable high pressure exists at lat. 30° on either side of the Equator. Considering this information along with the known patterns of easterly and westerly winds, Ferrel proposed that three circulation cells exist in each hemisphere.

A subtropical cell moves air up and away from the Equator to about lat. 30°, Ferrel said. There the cooled air sinks back to earth, creating the belt of surface high pressure that was so well documented in Maury's study. Some of this air, Ferrel said, then returns to the Equator to complete the cell described by Hadley.

But not all of the air returns to the Equator, Ferrel noted: Some of it moves toward the Poles at low altitudes and is gradually warmed by the surface. At about lat. 60°, where Anchorage, Alaska, and the South Sandwich Islands are situated, this poleward-moving air encounters dense, frigid air moving away from the Poles; it rises again and doubles back toward the 30° belt, completing a second circulation cell. A third cell, according to Ferrel, is formed by cold air circulating from the Poles to around lat. 60°, where it warms, rises and returns. The surface westerlies in the midlatitudes

occur, Ferrel said, because between lat. 30° and lat. 60° the air near the surface is moving poleward and is deflected toward the east in response to the Coriolis effect.

Ferrel also concluded that the Coriolis effect influences cyclones and anticyclones. In the Northern Hemisphere, air flowing in toward centers of low pressure curves to its right, creating the counterclockwise spin characteristic of cyclones. Moving outward, away from a high-pressure anticyclone, the air's curve to the right creates a clockwise spin. In the Southern Hemisphere, the situation is reversed: Cyclones turn clockwise and anticyclones counterclockwise.

One consequence of Ferrel's three-cell model was a clear picture of the mechanics of a phenomenon that had baffled and infuriated mariners for centuries—the infamous doldrums. A narrow, virtually windless zone cursed by the crews of countless sailing vessels, the doldrums remain stupefyingly still because the high heat at the Equator creates a chimney-like effect, drawing the returning northeasterlies and southeasterlies upward un-

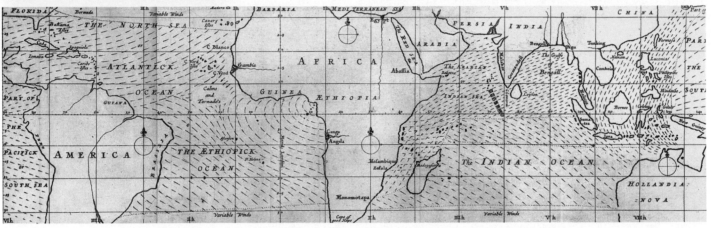

In 1686, astronomer Edmond Halley offered an explanation of trade-wind movement and a meteorological map (above) to illustrate the theory. He correctly deduced that winds near the Equator follow a reliable pattern because of the rising of sun-heated tropical air.

til they meet thousands of feet above the sea. Becalmed on glassy waters, hapless sailors often drifted for days in the blazing heat.

"Day after day the sea lay stagnant all around us," recalled seaman and author Alan Villiers when describing his passage through the doldrums on a four-masted windjammer in 1919. "The sails slatted and banged, useless and fretful, against the great steel masts, while the mariners hauled the ponderous yards about to every catspaw that came whispering deceitfully across the oily sea. None grew into usable wind. Some days we moved not at all; on others, backward." Often, when the moist air in the doldrums rises, it cools rapidly aloft, loosing torrents of rain through towering thunderheads. "The rain beat down unceasingly," another windjammer sailor recalled; "the air was steamy and left us gasping like landed fish."

Skilled captains did their best to find the shortest path through the doldrums, but even the most experienced could go wrong, since the doldrums do not stay in one place: The zone moves with the seasons—slightly northward during the northern summer and back toward the south in the opposite season, when the Northern Hemisphere is tilted away from the sun. The midpoint of this perennial drift is not, however, exactly on the Equator. Because the Northern Hemisphere contains more land, which heats up faster than water, the doldrums tend to be found slightly north of the line.

From the torrid doldrums, the air rises some 10 to 11 miles through the layer of air called the troposphere (*pages 62-63*) until it reaches the tropopause—the boundary between the troposphere and the stratosphere. Because the temperature stops decreasing with altitude at the tropopause, the air rising from the earth halts at this ceiling and, like water spewed from a vertical fountain, spreads outward to the north or south (very little wind crosses the Equator into the neighboring hemisphere). As the poleward-moving air cools and sinks some 30,000 feet toward earth, it is compressed, and the increased pressure reheats it, evaporating any remaining moisture before the air reaches the surface. As a result, the belts of high pressure that girdle the earth near the latitudes of northern and southern Africa and southern Australia are also hot and dry. On land, these desiccating high-pressure zones account for the world's great deserts—the Sahara and the Arabian Desert in Africa, the Mojave Desert in the United States and the Great Victoria Desert in Australia. At sea, the hot, descending air creates an infernal stillness.

In the North Atlantic, this dry, high-pressure zone is called the horse latitudes, because horses being ferried from Europe to the New World often perished of hunger and thirst as sailing ships languished in the blazing calm. According to nautical lore, their carcasses were tossed overboard and this grim litter gave the area its nickname. "Dead calm and hot as hell," one sailor wrote of this realm. "The deck oozes tar from between the planks. The iron parts of the ship are too hot to touch. There's not a ripple on the water, not a fleck of cloud in the sky. The sun has all but dried the sweat out of the crew." Eventually, the air does move out of the horse latitudes back toward the Equator to complete the great circulation cell, which is called a Hadley cell.

A similar pattern of circulation exists over the polar regions. There, the primary influence on the winds is the intense cold that persists because the snow-clad earth reflects almost all of the sun's scant heat back into space. At the top and bottom of the world, the frigid air continually sinks, pushing the surface air outward. Temperature differences within the polar regions are generally not very great, so these winds are, on the average, relatively weak, despite the occasional gale produced by local variations. Since cold air cannot hold as much moisture as warm air, polar winds also tend to be dry, and snowfall is light. The Poles are covered with thousands of feet of snow and ice because the snow that does fall remains for decades, even centuries—virtually none of it melts. As the cold polar air moves down into the warmer latitudes, the Coriolis force takes hold, bending it to create the prevailing polar easterlies, just as Equator-bound winds in the Hadley cell curve, becoming easterlies. The polar air meets the warm air of the Temperate Zone near lat. 60°. Where these two great air masses collide, the battleground is marbled with titanic swirls and eddies—the great cyclones and anticyclones of the middle latitudes.

In the temperate latitudes, the average annual temperatures are relatively mild, but the great temperature and pressure differences between the subtropical and the polar air masses give rise to powerful winds. Large land masses affect the circulation in the Northern Hemisphere and reduce surface wind speeds somewhat, but in the Southern Hemisphere, enormous expanses of open ocean offer little constraint. In the South Pacific, where clear ocean spans more than 5,000 miles, the westerly winds build to fearsome strength. Frequent 100-mile-per-hour winds create 60-foot waves, whose frothy visages were called "graybeards" by sailors fighting for their lives on ships rounding Cape Horn. From the bitter experiences of seamen battered by the midlatitude southern westerlies came the titles "Roaring Forties," "Howling Fifties" and "Screeching Sixties." One forlorn spot on the Antarctic Circle, Adélie Land on the coast of Antarctica, suffers wind speeds that average 50 miles per hour.

Overall, the winds of the temperate latitudes tend to be westerlies, but they are so changeable that the simple cell model, which is quite satisfactory when applied to the equatorial and polar zones, is an inadequate explanation for midlatitude winds. There is some cell-like circulation, but early in the 20th Century, meteorologists became increasingly aware that a great deal of the circulation in the Temperate Zone is lateral, rather than vertical, in nature. For the most part, the flow takes place around the great low-pressure cyclones and high-pressure anticyclones that form a steady progression across the temperate latitudes.

Air moves from regions of high pressure, where the air is dense, toward regions of low pressure; the force responsible for this movement is called the horizontal pressure-gradient force. As the air moves "downhill" along the gradient, it is deflected by the Coriolis force—to the right in the Northern Hemisphere, the left in the Southern. Eventually the pressure-gradient force, trying to shift the air from high to low pressure, and the Coriolis force, turning it to one side, balance each other: The air curves on its way down the pressure gradient until it is flowing at right angles to the path between the pressure systems.

This equilibrium of the two forces—called the geostrophic, or earth-turning, balance—accounts not only for the circulation patterns of cyclones and anticyclones that Ferrel defined, but also for the forces driving the winds in and around the vortexes. In the Northern Hemisphere, the surface air flowing outward from a high-pressure system in response to the pressure-gradient force curves to the right, obeying the Coriolis force, and in seeking a geostrophic balance spirals outward in a clockwise circulation. Air drawn into a low-pressure center by the pressure-gradient force is also deflected to the right by the Coriolis force: The compromise is a counterclockwise flow around the low.

While much of the wind activity in the middle latitudes takes place in the lateral wheeling of air around cyclones and anticyclones, another major factor is the pressure gradient caused by the temperature differences between tropical and polar air. The gradient's slope occurs because in a column of cold polar air, the air molecules are pressed down toward the ground to form a high-pressure system at the earth's surface. Higher in the column, fewer air molecules remain and the air pressure decreases steadily with height. On the other hand, the air molecules in a column of warm, tropical air are not compressed as tightly, so the pressure does not decrease as rapidly with altitude. Therefore, at higher elevations, the air-pressure differences between tropical and polar air columns are always greater, and the pressure gradient is always steeper, than at the surface. The pressure gradient for an entire hemisphere at elevations of approximately 30,000 feet tends to tilt downward toward the cold Pole. The Coriolis force bends the resulting flow of air toward the east, creating the prevailing westerly winds in the midlatitudes.

This general pattern is, however, anything but uniform in intensity. In summer, when the heat is more evenly distributed between the Equator and the Poles, the air-pressure gradients are relatively slight. But in winter, the dramatic differences in the temperatures of air masses create steep pressure gradients and generate high winds. In addition, irregularities in the earth's surface features affect the winds. Mountains, valleys, deserts, forests and the great bodies of water all play a part in determining how the wind blows, especially in the middle latitudes.

The direction of the Temperate Zone winds is most often altered by gigantic eddies—the cyclones and anticyclones that are embedded in the general westerly flow. Scientists working in Bergen, Norway, after World War I recognized that these great high- and low-pressure areas chase each other around the globe in the Temperate Zone, along the border between the temperate and polar air masses. The Bergen scientists explained for the first time how the collision of warm and cold air masses, with their countervailing wind systems, spawned disturbances in a constant progression from west to east. A member of the Norwegian team, Carl-Gustaf Rossby, later developed a theory that accounted for this complex chain of events.

In the early 1950s, scientists at the University of Chicago devised a simple experiment, using an ordinary dishpan on a turntable, that provided a graphic demonstration of Rossby's theory. Fitting the dishpan with apparatus that cooled its center and heated its rim, they poured into it an inch or so of water. They dropped in some aluminum powder to show how the surface of the fluid behaved and some dye to show motion below the surface. Then they mounted an overhead camera that rotated with the turntable to record the movement of the particles in the water. The turntable's rotation simulated the earth's rotation, and the water the circulating atmosphere. The dishpan's cold center represented one of the Poles, and its rim the warm tropics.

When the dishpan was stationary, the metal particles moved by convection upward from the heated edge and inward toward the cold center, then sank and returned to the rim, forming a classic Hadley cell. When the pan was rotated slowly, the particles moved neatly and evenly around, deflecting in response to the Coriolis force in a manner that was still similar to the Hadley circulation model. But when the speed of rotation was doubled to make it more analogous to the rate of the earth's rotation, the picture changed dramatically. The circulation cell marked by the particles remained much the same, but midway between the dishpan's edge and center—the area representing the midlatitudes—the particles formed a wavy line roughly tracing the circumference of the dishpan. Dye revealed horizontal whirls and eddies—corresponding to the cyclones and anticyclones of the atmosphere—below the wavy line.

Thus it was confirmed that fluids have a natural tendency toward instability in a middle zone between temperature extremes on a rotating body such as the earth. This instability is the source of the cyclones and anticyclones, which, it is now known, accomplish much of the work of transporting heat from the tropics to the Poles. The zonal pressure gradient that tilts toward the Poles at high altitudes and the turbulence that follows from the general laws of fluid dynamics are far more significant in the midlatitudes than Ferrel's vertical cell model.

There was, however, another curious element visible in the photographs of the dishpan experiment. Along the wavy middle-zone pattern of circula-

Standing Watch in a Perennial Purgatory

Even though it is a mere foothill in comparison with the world's greatest mountains, 6,288-foot Mount Washington in central New Hampshire is regularly buffeted by some of the worst weather on earth. Global patterns of atmospheric circulation and the local topography combine to account for the punishment that it receives.

Major storms generally travel across the United States along one of three primary tracks, all of which converge near Mount Washington; and the site gets the maximum effect from each storm because—as is the case with any mountain—the wind accelerates as it passes over the peak. The weather observatory that was established on Mount Washington in 1934 to study such meteorological extremes registered the world's record gust—231 miles per hour—in the same year.

The crew of the observatory confronts ferocious weather almost daily. In winter, when winds frequently exceed 100 miles per hour, temperatures drop to −30° F. and dense fog shrouds the mountaintop, the men must tether themselves to ropes as they move between buildings. "This work appeals most to those who love nature," according to one member of the crew, "and to those who have a reclusive bent."

Streamers of rime ice—cloud droplets
that freeze on impact with stationary objects—
feather the upwind side of a fence post
and a wind turbine erected for testing in Mount
Washington's mighty winds.

Hunched against the wind, a crew member
makes the rounds of Mount Washington's
instrument buildings. Readings are reported to
the National Weather Service every three hours.

tion, there appeared a heavier collection of streaks that looked like a kind of river winding around the pan. It threaded its way between the areas of turbulence and changed its course frequently, but it was always there. The scientists immediately connected this peculiar manifestation with a strange wind that had been encountered a few years earlier by World War II bomber pilots.

On November 24, 1944, ninety-four B-29 Superfortress bombers were approaching Tokyo on the first mass bombing assault of the War on the capital of the Japanese Empire. At altitudes between 27,000 and 33,000 feet, the planes made a turn to the east over Mount Fuji and began their bombing run. Suddenly, the startled pilots found themselves roaring past landmarks at a ground speed of almost 450 miles per hour, about 90 miles per hour faster than the theoretical top speed of the aircraft. By the time most of the bombardiers found their targets, it was too late to make allowances for the increased velocity of the aircraft; of the more than 1,000 bombs dropped, only 48 fell anywhere near their objectives. By military standards, the mission was a dismal failure, but meteorologists were fasci-

High-altitude clouds over the Nile River
valley trace the path of a powerful jet stream.
In the inset picture, similar jet-stream-
sculpted clouds stripe a Florida sky.

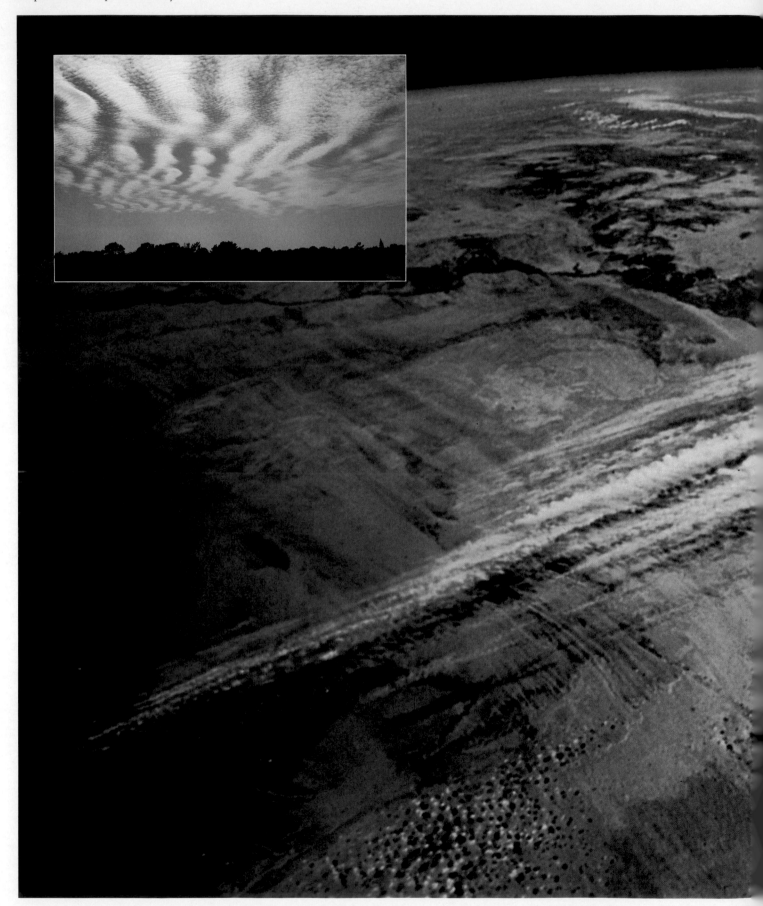

78

Sheets of monsoonborne rain drench villagers on Borneo. The seasonal monsoons deposit some 120 inches of rain on the island each year, sustaining one of the world's great rain forests.

nated by the discovery of a river of air aloft that was rampaging eastward at some 140 miles per hour.

After the War, scientific probings with high-altitude aircraft and balloon flights confirmed the existence of ribbons of rapidly moving air, normally 180 to 300 miles wide and up to two miles thick, that meander around the earth at altitudes of 30,000 to 45,000 feet. The average speed of the wind in these jet streams, as they came to be known, is between 60 and 115 miles per hour, and speeds exceeding 290 miles per hour have been recorded. In addition to the jet stream discovered by the bombers over Japan, others were found over the subtropics and high in the stratosphere over the polar regions.

Further investigation revealed that the streams occur where changes in temperature and air pressure—the ultimate instigators of wind flow—are most abrupt, usually near the tropopause along the boundaries between hemispheric temperature zones. It became increasingly clear that the jet streams are a necessary part of the process of transferring heat energy quickly away from the Equator toward the Poles to attain thermal balance. A better understanding of their role in the process emerged in the 1950s, when, with the assistance of computers, scientists were able to construct mathematical models of the forces involved. The jet streams were seen in a new light—as a straightforward consequence of the law of the conservation of angular momentum.

This law of physics is most clearly demonstrated when a figure skater,

80

twirling slowly with arms extended, draws her arms inward and by so doing speeds her rotation dramatically. Roughly stated, the law is that a rotating body speeds up as it nears the center of rotation. Like the arms of a figure skater, air rotating with the earth at the Equator speeds up as it moves toward the Poles. In the absence of other forces, the speed would increase constantly until titanic winds perpetually lashed the polar areas.

But just as the figure skater eventually slows and stops, partly because of the friction of the skates against the ice, so the winds are slowed by friction, and their vast energies are partially converted into other forms— the great whirling eddies of the cyclones and anticyclones, vertical turbulence and the jet streams.

The two most important jet streams, blowing from west to east, are located in the upper reaches of the polar front—the border between the temperate and polar air masses. The strength and permanence of the jet streams along this wavy boundary are related to the abrupt drop in temperature in the vicinity of the polar front, which creates a steep pressure gradient aloft near the tropopause. High above the horse latitudes at the edge of the tropics flow the subtropical jet streams, also blowing from west to east. Typically, they are strongest in the winter, when the temperature difference between the tropical and temperate air zones is greatest. From time to time the polar-front and subtropical streams merge, and their combined power can generate severe storms below.

Three other intermittent jet streams race across the skies. Two are called polar night jets because they appear in the dark winter months, high above the North and South Poles. During the months when a polar region is engulfed in frigid darkness, the air high over the Poles becomes much colder than the air over the Equator. This difference in temperature gives rise to extreme air-pressure differences in the stratosphere, which, in conjunction with the Coriolis force, create the westerly polar night jets at altitudes of about 30 miles.

Another seasonal jet stream forms at the tropopause over the Indian Ocean and Africa during the summer. This reverse jet stream—so called because it flows from east to west—arises because the huge Asian land mass heats so fast that for a time the air above parts of it is warmer than air at the same altitude above the Equator. This reversal of the normal pressure gradient aloft accounts for both the reverse jet stream and, at the surface, India's famous monsoons.

The monsoon is one of the many winds of the world whose influence is so pronounced and regular that they have assumed their own identities. The name "monsoon" comes from the Arabic word *mausim,* meaning "season," and refers to the fact that this wind blows from one direction for six months, bringing bountiful, life-giving rains to the subcontinent, then reverses itself.

The annual cycle of the monsoon can be said to begin in the winter months, when air from the cold, dry interior of Asia north of the Himalayas flows toward a warm, low-pressure area over India. Turned westward by the Coriolis force, this wind blows all the way to western Africa. Then, slowly but inexorably, the Asian interior begins to heat up and the offshore wind dies down as the temperatures on land and at sea equalize. The hot, dry season reaches its peak in April. At this time, the skies over India are virtually cloudless and the fields are parched. Then in May, a modest breeze blowing from the southwest announces to the Indians that the summer monsoon is on the way.

Sucked toward the land by the low pressure beneath the heated and rising inland air, the monsoon clouds come in over the Malabar Coast and strike the Ghat mountain chain in southwest India. Forced upward by the mountains, the air cools and releases its moisture as rain—often more than 200 inches in just half a year. Sweeping northward, the monsoon crosses India, drenching fields and sometimes flooding villages. In the province of Assam at the foot of the Himalayas, the deluge is astounding; one small town,

A hot, dry wind known as a foehn or chinook can occur even in the wintertime when air crosses a mountain range. First the air is lifted and cooled until its moisture condenses; then, as the dried air sinks on the mountain's lee side, it is compressed and heated.

MOIST AIR

HOT, DRY AIR

Cherrapunji, endures an average annual rainfall of 425 inches, and once received a total of 1,042 inches in one year, a world record.

In September or October, as the temperatures over the land and sea once more even out, the southwest wind dies down, and the rains slacken and finally cease. As autumn progresses, the continental interior north of the Himalayas cools, and the wind again begins to blow from the northeast.

This exchange of moist sea air and dry land winds is duplicated on a smaller scale almost every day along coastlines throughout the world. The phenomenon is that of the sea breeze. Like the monsoon, the sea breeze is caused by the different rates at which land and water absorb and retain the sun's heat. During the day, the land heats more quickly than the water, and the air over the land rises, creating low pressure at ground level. The sea air flows in across the beach toward this low, while the air above the land flows back out to sea at high altitude, completing the cell-like circulation. At night, the situation is reversed: The land cools more quickly than the water, and as the warmer sea air rises, the wind blows off-shore to replace it.

The sea-breeze cycle is not confined to seashores. Any large body of water is likely to have these cyclical breezes, although to a lesser extent. Chicago's eastern strip along Lake Michigan has been known to cool as much as 18° F. in one hour as a result of the wind blowing off the lake. In the early spring, when the water is colder than the land, moisture-laden lake winds bring bone-chilling cold and often heavy snow.

Far inland, daily temperature changes create a similar cycle in mountain valleys. Striking the elevated terrain, the morning sun heats the slopes, causing the air over them to rise. The air above the valley sinks to replace the air flowing upslope, creating a circular flow much like that at the shore or in a Hadley cell. At sundown, the mountain air cools and, assisted by gravity, the wind reverses and pours down the slopes. In many cases, the overall change from the upslope to the downslope wind occurs at about the same time every day; residents in the Swiss Alps find that they can almost set their watches by the daily wind shift.

Mountains, the greatest natural wind barriers, are responsible for some of the strangest winds on earth. Among the oddest is the foehn (pronounced like "fern" without the r), which often roars through Alpine valleys, particularly in the late winter and autumn. The foehn is a hot, dry wind that seems to defy logic by moving down from snow-clad summits. To the Swiss it can be a blessing, for it ripens fruit in the fall and enables farmers in the Rhine Valley to raise corn and grapes.

The process that engenders the foehn begins when air rises from lowlands to mountain peaks. As the air moves up the mountains, it cools and its moisture precipitates out. After crossing the summit, the air spills into the valleys, and as it descends it becomes compressed. This compression of air molecules heats the air at the rate of some 5.5° F. every 1,000 feet. After falling some 5,000 feet to the valley floor, the wind is almost 30° F. warmer than it was when it crossed the mountain. The Swiss call the springtime manifestation of this hot, dry air "the Snow-eater," because it can wipe out most winter drifts in one day.

Now and then, however, the foehn can be dangerous. Its hot breath dries out foliage quickly, increasing the likelihood of forest fires, and it may melt snow so rapidly that severe flooding results. The onset of such a calamity was described in detail by a 19th Century tourist. "In the distance is heard the rustling of the forests on the mountains. The roar of the mountain torrents, which are filled with an unusual amount of water from the melting snows, is heard afar through the peaceful night. A restless activity seems to be developing everywhere, and to be coming nearer and nearer. A few brief gusts announce the arrival of the foehn. These gusts are cold and raw at first, especially in winter, when the wind has crossed vast fields of snow. Then there is a sudden calm, and all at once the hot blast of the foehn bursts into the valley with tremendous violence, often attaining the velocity of a gale which lasts two or three days with more or less intensity, bringing confusion everywhere; snapping off trees; loosening masses of rock; filling up the mountain torrents; unroofing houses and barns—a terror in the land."

A similar—and no less hazardous—wind of the foehn type is Southern California's celebrated Santa Ana, which emerges from the sun-baked Great Basin between the Sierras and the Rocky Mountains and pours over into the Los Angeles basin. Hot and dry, it plunges into the heavily populated valleys and canyons. The Santa Ana's gusts, as high as 68 miles per hour,

Ceaseless Cycles of Atmospheric Flow

Fickle as the winds may seem, they are marvels of predictability when considered on a global scale. Their movements are energized by heat: Warm, buoyant air rises, and cold, dense air flows in to replace it—creating wind. Driven by the sun and shaped by the spinning of the globe, the wind struggles eternally to mediate between the steamy tropics and the frigid Poles.

The process can be said to begin at the Equator, where heated air rises over the doldrums, a region of light, variable surface winds. As it rises, the air cools, moves outward, and eventually becomes so dense that it sinks, forming belts of high pressure and relative calm—known at sea as the horse latitudes—about 30° north and south of the Equator. This dense air is drawn inward again to the Equator, completing a circular pattern of flow known as a cell. Deflected by the Coriolis effect of the earth's rotation, the air near the surface moving toward the doldrums becomes the easterly trade winds.

Some of the air from the tropical cells is transferred to two weaker circulation cells over the temperate latitudes. The flow in these midlatitude cells is the reverse of that in the tropical cells (and also of that in another set of cells nearer the Poles). Thus the air near the surface at midlatitudes moves poleward and is deflected by the Coriolis effect into prevailing westerlies.

Along the polar front—the constantly shifting zone of contact between the midlatitude westerlies and the cold easterlies of the polar air masses—temperature and pressure differences are often extreme. Here are found the great whirling cyclones and high-altitude polar jet streams that engender some of the world's mightiest winds.

A map based on long-term averages shows the patterns of the world's prevailing winds. The cross-sectional diagrams indicate the position and shape of the rotating circulation cells, and the location of the jet streams. The cross section of the polar cells, however, is distorted by the whole-earth projection used here.

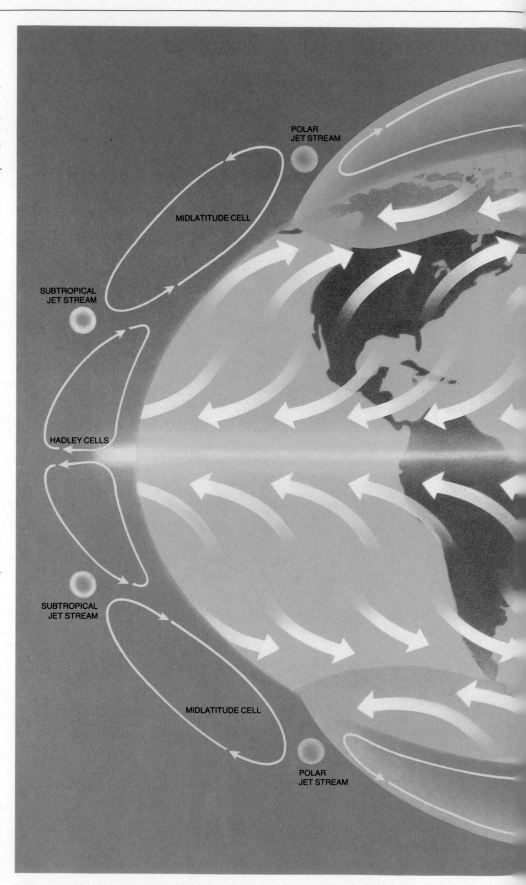

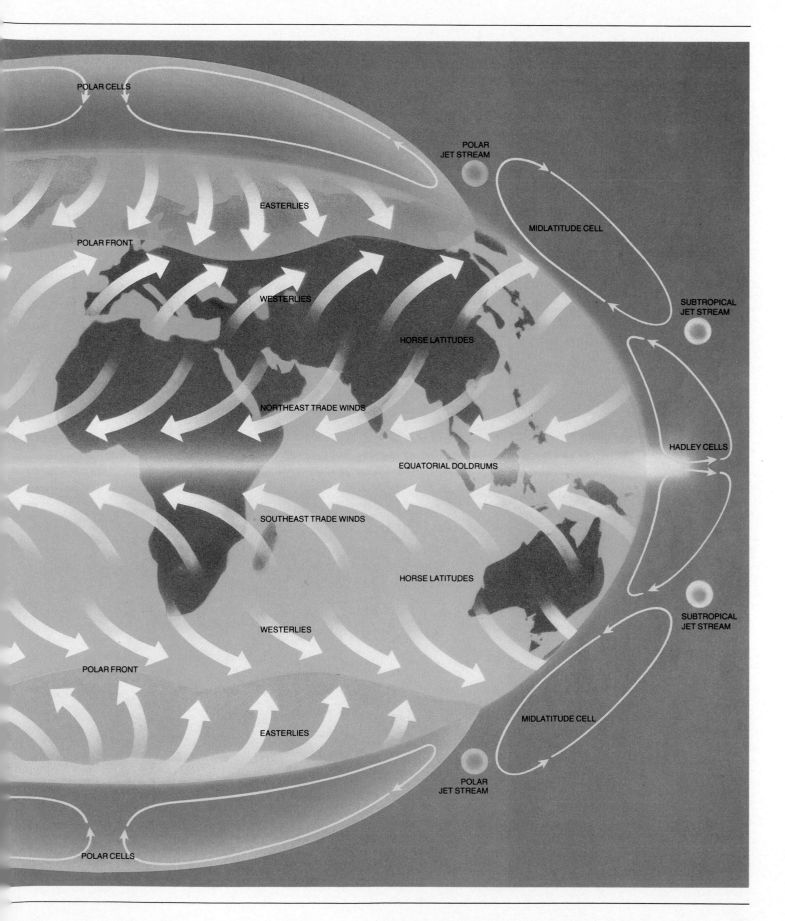

carry small stones and bits of gravel that can pierce windowpanes like bird shot, leaving neat little holes. Although the scorching wind may break a winter cold snap, it can also suck all the moisture out of the chaparral, buckthorn and sage that line the canyons. Then the slightest spark can start a rampaging wildfire, to be fed with copious amounts of oxygen by the furious Santa Ana. Such conflagrations have caused many millions of dollars' worth of damage.

A more benign North American foehn delights the farmers and cattlemen of the northwestern plains of the United States and Canada. Called the chinook because the early settlers thought it came from a Chinook Indian village on the lower Columbia River, this wind is now known to originate in a moist wind from the Pacific Ocean that releases its moisture as precipitation over the Rockies, then is compressed and heated by its descent onto the frozen plains. The hot blasts of the chinook frequently rescue herds of cattle from starvation by uncovering grasses locked beneath the ice and snow. One day in 1900, a chinook raised the temperature in a Montana town by 31° F. in three minutes, and chinooks have been known to devour a 10-inch snow cover overnight. Often they remove snow without melting it, through a process known as sublimation: The snow simply vaporizes, vanishing into the air.

Some foehn-type winds are so cold at the outset that the warming effect of compression is barely noticeable. Such a wind is the mistral, which originates in a high-pressure area over central France and blows down the Rhone Valley toward the Mediterranean coast. Stormy, cold and dry, it can knock down chimneys and overturn railroad cars.

No less dramatic than the mountain winds are those that roar across the desert wastelands of the world. Few who endure such searing blasts would be comforted by the knowledge that these winds are nature's way of distributing the desert's extreme heat quickly. Because deserts are typically found where hot, dry air is sinking in an anticyclone, the air in the high-pressure centers flows outward quickly toward the nearest low-pressure center. By any measure, the most dreaded desert wind is the sirocco of the Sahara. When attracted by a low over the Mediterranean Sea, the sirocco, blowing unimpeded across the open desert, can spread misery throughout northern Africa.

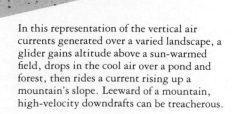

RISING CURRENT

DOWNDF

COOL, SINKING AIR

WARM, RISING AIR

In this representation of the vertical air currents generated over a varied landscape, a glider gains altitude above a sun-warmed field, drops in the cool air over a pond and forest, then rides a current rising up a mountain's slope. Leeward of a mountain, high-velocity downdrafts can be treacherous.

"The leaves of the trees folded before the eye," wrote one Frenchman witnessing the phenomenon in Algiers in the 1860s. "The clouds of flying sand soon eclipsed the obscured disk of the sun; and the different shades of yellow, orange, saffron, and lemon color melted into a mass of copperish color impossible to describe."

One British mining engineer, caught in the Abyssinian desert when the sirocco hit, later described his unnerving experience. "The sandhills were displaced under our very eyes; all that we could see about us was in motion. There were seen waves, breakers that rolled onward like wind-blown water. The sand was whipped so fast from their crests that within a few moments the troughs under them became filled, and new waves rose up where they had been. The process was continuous, and the effect was that of a slow-rolling sea."

The sirocco frequently crosses the Mediterranean to bring discomfort to Spain, Italy and Greece—though in transit it changes character. Picking up moisture from the sea, it arrives in southern Europe laden with humidity. An epidemic of severe headaches may ensue, and in Sicily and southern Italy crimes of violence are said to increase dramatically. Most of the wind's power is broken by the time it reaches the Pyrenees and the Alps, but from time to time it presents northern Europeans with a special and unwelcome gift: innumerable particles of Saharan sand. In one year, 1901, an estimated two million tons of the ocher-colored dust was deposited over the Continent. Once it caused a reddish snow to fall on 16,000 square miles of Germany, and in the year 582 A.D., a dust-tinted rain falling on Paris was thought to be a "shower of blood."

Not many winds have such bizarre consequences, but because the atmosphere is as integral a part of the planet as the land and seas, the wind will always have a profound effect on all earthly life. Every breeze in the world, no matter how small, has the potential to build into something larger. The warming of a hillside, the shadow cast by a building, the cooling of a pond, the drying out of a sandbank—any of these set the restless air in motion. Even the sunny side of a pebble has its own minuscule updraft, which, if conditions are right, can combine with thousands of other updrafts to make a massive body of turbulence that may well chart the day's course for the great atmospheric wind machine. **Ω**

On some of the rolling hills of the ranch
country near California's Diablo Mountains, there bristles what at first glance
appears to be a startling new crop: a
field of 40-to-60-foot-high windmills—
more accurately called wind turbines—
whirling like pinwheels in the breeze.
This forest of three-bladed wind ma-
chines is part of the Farrell/O'Keefe
Wind Farm, located in Altamont Pass,
through which the surrounding mountains funnel a strong, steady wind averaging about 15 miles per hour. Thus
powered, each of the turbines pumps
about 100,000 kilowatt-hours of electricity into the local utility's power grid
each year. The wind intensifies in the
summer, coinciding fortuitously with
the peak demand for electricity.

Such methods of harvesting the power
of the wind—the world's chief source of
energy prior to the Industrial Revolution—are enjoying a widespread renaissance as a power-hungry world looks
for alternatives to fossil fuels. The ener-
gizing of an ancient concept with contemporary technology and materials is
producing a variety of hybrid mechanisms significantly different from the
colorful but plodding ships and windmills of history.

At Farrell/O'Keefe, for instance, a
computer monitors the 30-foot fiber-
glass blades as they whirl atop the sleek
steel towers; at any sign of malfunction
the computer turns the affected turbine
out of the wind, shutting it down until
it can be repaired.

Such sophisticated innovations may
increase the efficiency of wind turbines,
but they also increase cost. Until designs
are perfected so that turbines can be
mass-produced, it will remain difficult
for wind farms to turn a profit. But
Farrell/O'Keefe and other wind farms
cropping up on hills across the country
are speeding the arrival of a new technology of the wind.

An array of turbines in Altamont Pass,
California, draws power from the wind without
disturbing nearby cattle. Other designs
being tested (inset) include a vertical-axis rotor
(left) and a giant 300-foot blade.

Reinventing the Venerable Windjammer

Hardly the graceful clipper ship of yore, the wind-assisted Japanese oil tanker pictured at right may herald a new era for commercial sailing. When sensors aboard the vessel detect favorable winds, a computer slows the ship's engines and unfurls two 1,000-square-foot plastic sails, trimming them constantly in response to wind changes. The assistance is expected to cut the 236-foot tanker's fuel consumption approximately in half. The world's 25,000 major cargo ships consume an estimated four million barrels of oil per day: If harnessing the wind were to reduce consumption of oil by just 10 per cent, the economic impact would be enormous.

Continuing experiments with sail and hull design, the application of modern synthetic sail materials and the use of satellite-assisted meteorology and navigation hold out the hope that the most efficient use of wind power in sailing's 5,000-year history may be at hand.

The oil tanker *Shin Aitoku Maru* cruises under sail near Japan. Rigid plastic sails, shown unfurling below, were found in wind-tunnel tests to be more efficient than fabric sheets.

Blueprints for Culling Aerial Energy

The challenge of efficiently converting the boundless energy of the wind to electrical power has captured the imagination of engineers, professors and backyard tinkerers around the world. The designs that are shown here address the primary difficulty of harnessing wind power—its diffuse, skittish and unreliable nature.

Below, a shroud surrounding a three-bladed wind turbine increases efficiency by accelerating the movement of air past the blades. The flying wind turbines at right are designed to reach the jet-stream winds that roar along at speeds frequently exceeding 100 miles per hour at altitudes of 30,000 feet.

The far more complex design illustrated at far right would emulate nature's most powerful windstorm—the tornado. Its array of vents is intended to draw wind from any direction and manipulate the flow into a maelstrom that could theoretically yield 35 times more electrical energy than unassisted turbine blades.

If such ideas as these could effect an increase in efficiency of just 5 per cent over existing systems, the wind might be economically harnessed.

Propeller-like turbines mounted on tethered gliders (*above*) would exploit the jet stream to generate electricity. Tilted up, the turbines would provide lift to help the gliders ascend.

Wind entering the shroud that encircles the rotor blades of a diffuser turbine (*left*) is first funneled to the blades, then allowed to expand as the shroud widens. The resulting drop in air pressure behind the whirling blades draws more air into the turbine.

In a tornado turbine (*right*), wind entering vertical slots in a 1,800-foot tower would be whirled around the interior, forming a low-pressure core that would suck air up through the horizontal turbine blades at its base.

Chapter 4

THE COUNTENANCE OF THE CLOUDS

In that extraordinary intellectual era at the beginning of the 19th Century when so many well-to-do persons took pleasure in dabbling in the natural sciences, no one was more intrigued with the world around him than a young London apothecary named Luke Howard. The son of a wealthy manufacturer, Howard spent his working hours operating a chemist's shop and developing what became a substantial chemical manufacturing business. While he was dedicated to his profession, he was at the same time enthralled by his avocation—scanning the skies and thinking about what he saw there. The behavior of the atmosphere was a subject on which he could be positively lyrical.

"The *sky* too belongs to the Landscape," he once wrote. "The ocean of air in which we live and move, in which the bolt of heaven is forged, and the fructifying rain condensed, can never be to the zealous Naturalist a subject of tame and unfeeling contemplation." But Howard brought more than a sense of wonder to his hobby. An astute observer, he made a contribution to the fledgling science of meteorology that was so perceptive and all-encompassing that it became, and remains to this day, part of the language of the science.

Like many other educated men of his day, Howard belonged to a small club whose members presented to one another inquiring papers on scientific subjects. His particular circle was called the Askesian Society, a name derived from the Greek word for "self-improvement" and "discipline." Meeting in London during the winter of 1802-1803, the group heard the 30-year-old apothecary read a paper on the subject of clouds. Investigation into the "countenance of the sky" was badly in need of some kind of system, he said; farmers and mariners understood these "aggregates of minute drops," but there was no way to record and discuss their expertise. His observations had persuaded him that clouds could be classified according to their appearance and behavior, and he proposed to give names to "such of them as are worthy of notice." The labels would be in Latin, a universally accepted form of scholarship. It was also, Howard noted later, a language that had been drilled into him from boyhood. "I have acquired more Latin than I have since been able, much neglect of study notwithstanding, to forget."

All clouds belong to one of three distinct groups, he said. The wispy, high-level formations, "first indicated by a few threads pencilled, as it were, on the sky," he named cirrus clouds, from the Latin word for "a lock of hair." The lumpy, individual clouds found nearer the ground, he called cumulus, meaning "heap." And the horizontal blankets of cloud, including

Altocumulus roll clouds, the result of continuous wavelike motions in a layer of moist air, band an early-spring sky. Water vapor cools and condenses when the air rises, forming a strip of cloud; as the air sinks, it warms and clears before being wafted upward again.

95

fogs, that usually cover wide areas were named stratus, from the Latin word meaning "widespread" or "layered."

He also combined the terms to describe cloud combinations. For example, when cumulus clouds became so crowded in the sky that they formed a layer, the resulting formation was dubbed cumulostratus. To denote "a cloud in the act of condensation into rain, hail or snow," Howard used the Latin word for "shower"—nimbus.

Although scientists in later years added several more names and prefixes to further delineate cloud formations, Howard's system, admirable for its simplicity, has remained enormously useful. He included in his study some speculation about how clouds form, but was unable to add much to the limited knowledge of the era about the behavior of air.

What has been learned since then has confirmed the intuitive notion shared by Howard and other early investigators that these evanescent, beautiful and strange denizens of the sky are vital to life on earth—and not only because they are a necessary precursor to nourishing rains. It is now clear that all clouds, even the relatively tiny cumulus, play an essential role in the constant exchanges of heat through the global atmosphere that moderate the earth's temperature.

Clouds blanket substantial portions of the planet every minute of every day and are never at rest. The towering majesty of anvil-topped thunderheads, the gloom of an impending blizzard and the bright, dainty puffs of fair-weather cumulus are but a few of their many manifestations. Yet for all their variety and their pervasive role in the world's weather, these airborne bodies of moisture represent at any one time an astonishingly minute portion of the earth's water—$\frac{1}{1,000}$ of 1 per cent. If all the water vapor in these countless whorls and clumps and layers of cloud, along with the invisible water vapor in the atmosphere, were somehow condensed and evenly distributed on the surface of the earth, the result would be about an inch of rain worldwide.

A pharmacist by profession, Luke Howard was a meteorologist by lifelong avocation. He is considered the father of British meteorology, primarily for his 1803 classification of clouds and his theories about weather.

The Ionian philosophers of Asia Minor, who were among the first to develop theories about the atmosphere, believed in the Eighth Century B.C. that clouds were a thickened form of wet air. Variations on that theme persisted even after the sagacious Frenchman René Descartes first declared in the 17th Century that air and water vapor are two different things. A century later, chemists were avidly analyzing all manner of solid substances, but little more had been learned about the chemistry of the air. Then, in 1751, a French physician named Charles Le Roy made a highly significant observation about its behavior.

By sealing damp air in a glass container and observing carefully while it cooled, Le Roy found that dew appeared on the inside of the glass at a certain temperature; when he heated the vessel above that temperature the dew disappeared, and when the container cooled again, the dew reappeared. He had discovered a cardinal principle—that any parcel of air has a specific temperature, the dew point, at which the water vapor it contains will condense to liquid form.

As Le Roy repeated his experiment outdoors and studied the effect of various weather conditions on condensation, he found that the humidity, or moistness, of air is not an absolute quantity that can be satisfactorily described simply by determining the amount of water the air contains. Just as

a thick towel can absorb more water than a thin one without becoming
soaked, so air on a warm day, Le Roy reasoned, can hold more water vapor
and feel less humid than the same air on a winter day. A true measure of
air's humidity, he concluded, has to be relative; that is, it must compare the
amount of water vapor present in the air with the maximum amount of
water vapor that the air can contain.

Like many other pioneers of science, Le Roy did not fully develop the
useful concepts he had glimpsed, partly because his work was based on an
erroneous concept of how water vapor got into the air in the first place. He
thought it dissolved, as salt dissolves in water, and reappeared as a liquid in
the same way that salt can be precipitated out of solution. For this reason,
he used the word "precipitation" to describe rain, dew and snow. He clung
to the solution theory even though other experimenters had shown that
water can evaporate into a vacuum, thus raising the question of how something can dissolve into nothing. (Although the term "precipitation" turned
out to be chemically inaccurate, it has remained in the language.)

Another barrier to understanding at the time was the common miscon-
ception that water vapor consisted of hollow droplets. The notion had been
around since 1666, when an Italian priest, Urbano d'Aviso, wrote that
vapor was made up of "little bubbles of water filled with fire, which ascend
through the air as long as it is of greater specific gravity than they are; and
when they arrive at a place where the air is equally light, they stop." Even
the distinguished astronomer Edmond Halley thought along these lines,
and so stated in a paper delivered to the Royal Society seven years after his
celebrated proposal on the trade winds. All through the 18th Century the
theory claimed its adherents, few more eminent than the Swiss geologist
and physicist Horace Benedict de Saussure, the man who developed one of
the earliest instruments for measuring humidity. Saussure said that he had
actually seen the hollow droplets when he inspected the surface of a hot
cup of coffee with a powerful lens. "The lightness of these little spheres,"
he wrote in 1783, "their whiteness, their appearance, absolutely different
to that of solid globules, their perfect resemblance to the more volumi-
nous bubbles that are seen to float on the surface of the liquid, all leave no
doubt of their nature."

By the end of the 18th Century, experiments were beginning to show
that water vapor neither possesses hollow cores nor forms a chemical solu-

tion with air. In 1802 the English chemist John Dalton put forth the theory that water vapor is in fact a gas that behaves in the air more or less like any other gas—such as oxygen, nitrogen or carbon dioxide. Dalton, a Quaker who led a scholarly, reclusive life devoted to scientific research, wrote insightful treatises on many topics, including cloud formation, color blindness and the aurora borealis; eventually he proposed the first quantitative atomic theory, stating that all matter is composed of minute indestructible particles called atoms, which differ from one another only in mass.

Dalton wrote that water vapor is one of several gases that mix—but do not combine chemically in solution—to form air. From this precept came his famous law of partial pressures, which says that each of the air's many gaseous ingredients exerts pressure independently. The sum of these several pressures, he said, is the air-pressure reading obtained with a barometer. Within a few years, scientists were able to employ Dalton's law, as it came to be called, to formulate a more advanced concept of relative humidity.

Molecules of water vapor present in a parcel of air exert a pressure—the water-vapor pressure—that is independent of the pressure of the other gases in the parcel. If the air is relatively dry, its water-vapor pressure is low; higher water-vapor pressure outside the air parcel will move water-vapor molecules into the parcel until equilibrium is reached. If the air is over water, evaporating water molecules will continue to raise the vapor pressure until the air parcel is saturated. After the saturation point has been attained, water evaporating into the air displaces an equal amount of the vapor already present; often, the result is precipitation. Thus, Charles Le Roy's deductions about relative humidity were confirmed and explained by the laws of physics; relative humidity can be expressed as the ratio of the existing vapor pressure to the maximum water-vapor pressure possible—the saturation point—in a parcel of air.

Other investigators, including Dalton and Saussure, proved that temperature directly affects relative humidity. The warmer the air, the greater its capacity to hold water vapor and therefore the more moisture required to create high relative humidity. Conversely, if the air cools, the relative humidity increases and the existing moisture in the air approaches the saturation point even though the absolute moisture content is unchanged. For this reason, even in a heavy snowstorm, when the relative humidity is close to 100 per cent, the cold air may hold less than half the moisture contained in the same amount of hot desert air.

Continuing experiments and close observations of the actual behavior of water vapor generally supported this enlarged explanation, but an unavoidable problem remained. Water vapor can condense to liquid before the air reaches the saturation point, and even when saturation has been reached, condensation does not necessarily begin. Clearly there was another factor at work. Not until the last quarter of the 19th Century was it identified.

In 1875, a French scientist named Paul Jean Coulier began testing the properties of fog in a totally enclosed system, measuring its response to changes in pressure. He was puzzled by the fact that after a number of experiments, he could no longer get fog to appear in the container he had been using, even when the theoretical requirements of temperature, pressure and relative humidity were met. If he then let in some new air from outside the container, fog again formed. Coulier concluded that the missing ingredient, without which the fog would not appear, was dust.

The Many Phases of Precipitation

Moisture can spill to earth as a gentle, nurturing shower or a bruising, destructive hailstorm. But whether boon or curse, precipitation begins with the condensation of water molecules around tiny airborne particles called condensation nuclei; thousands of droplets must then combine into units large enough to fall to the ground without evaporating.

In the tropics, and some temperate-latitude regions, unusually large cloud droplets sometimes form. Falling rapidly through the cloud, they overtake smaller droplets and sweep them up, ultimately growing to raindrop size and spattering to earth.

But almost all precipitation outside the tropics, even summer rain, begins as ice crystals. These form in the upper reaches of thick cumulonimbus or nimbostratus clouds, where the temperature is below freezing but most of the cloud droplets remain a supercooled liquid. When a few of these droplets do freeze, the resulting ice crystals grow rapidly as water vapor freezes onto them. Eventually, they fall as snowflakes.

At first the snowflakes tend to clump together; the form in which they reach the ground depends on the temperatures they encounter as they drift earthward. Cold air preserves them and they arrive as soft snow; warmer air melts them and they fall as rain. When warm air is sandwiched between the frigid cloudtops and a region of subfreezing temperatures near the ground, the precipitation first melts, then either refreezes before it reaches the ground, yielding sleet, or freezes on impact to form an icy armor on branches and power lines.

Only hail follows a different pattern. The most widely accepted theory of its origin holds that strong thunderstorm updrafts repeatedly hurl a developing precipitation particle into the upper levels of the cloud. Each time, a new layer of moisture freezes onto the developing hailstone, until it gets so heavy that even powerful updrafts cannot sustain it and it falls to earth.

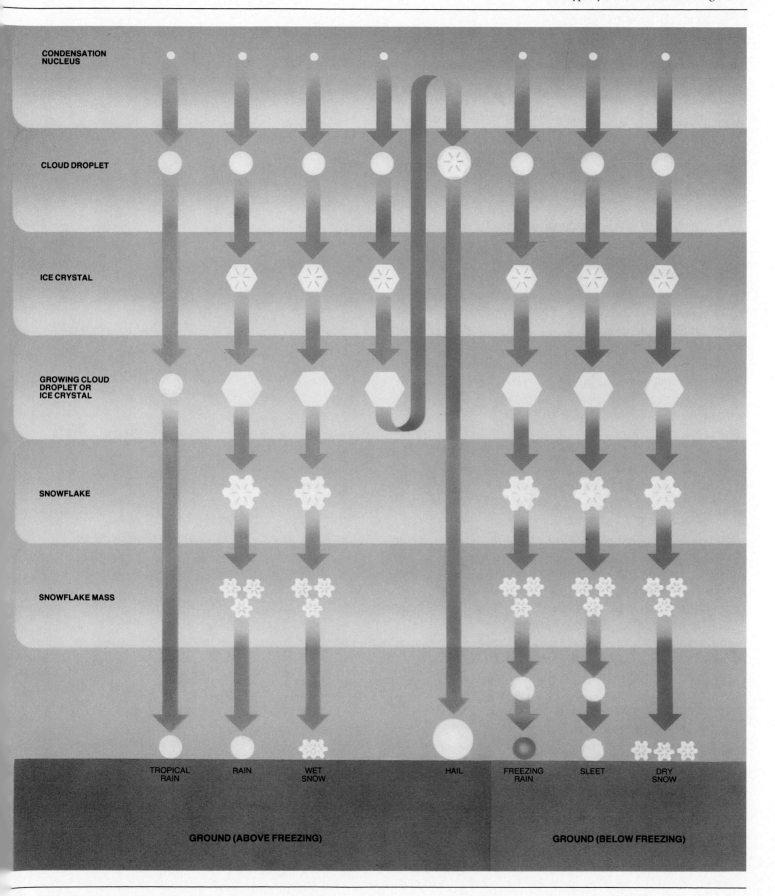

CONDENSATION NUCLEUS

CLOUD DROPLET

ICE CRYSTAL

GROWING CLOUD DROPLET OR ICE CRYSTAL

SNOWFLAKE

SNOWFLAKE MASS

TROPICAL RAIN   RAIN   WET SNOW   HAIL   FREEZING RAIN   SLEET   DRY SNOW

GROUND (ABOVE FREEZING)   GROUND (BELOW FREEZING)

Others confirmed Coulier's thesis that water vapor tends to condense around tiny particles of dust. The Scottish physicist and meteorologist John Aitken, after extensive investigations, identified the sources of the dust required for the formation of clouds and fog. "Everything in nature which tends to break up matter into minute parts will contribute its share," he wrote in 1881. "In all probability the spray from the ocean, after it is dried and nothing but a fine salt-dust left, is perhaps one of the most important sources." He also named volcanic dust, condensed gases and the residue of combustion. "From this it will be observed," he said, "that it is not the visible dust motes seen in the air that form the nuclei of fog and cloud particles. The fog and cloud nuclei are a much finer form of dust, are quite invisible, and though ever present in enormous quantities in our atmosphere, their effects are almost unobserved."

These particles are indeed infinitesimal. The smallest, called Aitken nuclei in honor of their investigator, are so tiny that as many as 5,000 of them could be jammed into one cubic centimeter. The larger ones are not much thicker than the film of a soap bubble, and even the so-called giants are only $\frac{1}{10}$ the width of a human hair. If no such particles existed in the air, clouds and rain would have a completely different character. The air would become supersaturated with vapor, without any clouds appearing; then, from time to time, massive clouds would form abruptly and ruinous downpours would lash the earth. Gentle rains would almost never occur.

Even with this new understanding that water droplets, and thus clouds, are encouraged to form by dust particles, one step in the hydrologic cycle remained unexplained: How do clouds become rain clouds—Luke Howard's nimbus category—with droplets large and heavy enough to fall to earth? Ordinary raindrops are 15 million times larger than a cloud droplet (and, incidentally, assume the shape of a tiny hamburger roll as they fall; the pear-shaped raindrops favored by cartoonists do not occur in nature). It had been assumed that the droplets simply grew, but no one knew enough about the actual sizes of raindrops or about their movements to be able to say for sure how they might form.

In 1911, German meteorologist Alfred Wegener—a brilliant thinker who would later help engender a revolution in geology by proposing that continents are in motion—came up with a theory that earlier researchers would probably have rejected out of hand. He suggested that almost all rain begins in the form of ice.

Two facts led Wegener to this conclusion: Water can be supercooled—that is, cooled well below its freezing temperature without turning to ice—and ice attracts water vapor. The attraction occurs because the vapor pressure near ice is lower than that near water; if ice crystals and water droplets are in close proximity in a moisture-laden air parcel, the faster-moving liquid water molecules will leave the droplets by evaporation and flow toward the lower vapor pressure—and the ice.

Wegener put these two pieces of evidence together in a theory of cloud physics that sketched in the last piece of the raindrop puzzle. Clouds that rise to great heights, he said, often contain both ice crystals and supercooled water droplets. Inevitably in such a mixture, he reasoned, the ice crystals grow at the expense of the water droplets until the crystals are heavy enough to fall toward warmer realms, where they may melt and become rain.

Wegener's idea remained just an interesting theory until, a decade later,

The Scottish chemist Joseph Black, portrayed in a posthumous caricature, lectures from a jumble of notes at the University of Edinburgh. In 1762, Black demonstrated the existence of latent heat, which is absorbed as water evaporates and is released again as vapor condenses.

a Swedish scientist happened to see it demonstrated. While vacationing in a Norwegian forest in 1922, Tor Bergeron, a member of the renowned Bergen school of meteorologists, noticed that fog behaved differently at different temperatures. When the temperature was above freezing, the fog descended all the way to the ground. But when the temperature dipped several degrees below 32° F., the air remained clear for a few feet above the ground. Recalling Wegener's theory, Bergeron decided that ice on the trees had indeed attracted and absorbed the fog droplets.

After a decade of extensive study, Bergeron announced in 1933 that every raindrop with a diameter greater than 500 microns—about the size of the droplets in a fine mist—originates as an ice particle. Such particles, he had discovered, begin to form only when cloud temperatures drop below 14° F.: Above that temperature they remain in the liquid state even when supercooled below 32° F. In the summertime, temperatures low enough to permit crystal formation occur at altitudes of two to three miles. To produce precipitation, clouds must either build to those heights or be seeded by ice crystals falling from higher cirrus clouds. In winter in the middle latitudes, the air in the troposphere is often cold enough to allow crystal formation at even lower levels. As long as they remain in subfreezing surroundings, the crystals grow rapidly by absorbing vapor from the supercooled water droplets, and then begin to fall, picking up more droplets along the way. In most cases, the crystals will combine to form snowflakes, which either arrive at the ground intact or melt on the way, becoming rain.

Although meteorologists now know there are exceptions, they still believe that this ice crystal process accounts for most of the world's precipitation in the middle latitudes, and they have further identified the details of the process. Even at subfreezing temperatures, the microscopic droplets are simply too small to freeze. In order to create a crystal structure,

Dry air warmed by the ground rises and cools at a steady rate *(right)*. If the rising air is moist, however, water vapor soon condenses into clouds, releasing latent heat *(far right)*. The heat slows the cooling of the moist air, maintaining its buoyancy and contributing to the upward growth of the cloud.

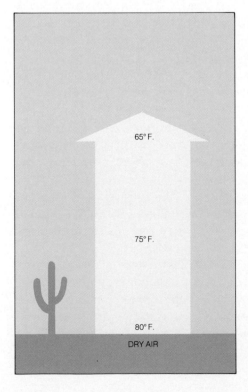

65° F.

75° F.

80° F.

DRY AIR

65° F.

70° F.

75° F.

80° F.

MOIST AIR

101

liquid water molecules must line up in a certain order. Most droplets are so small and contain so few free-floating liquid molecules that there is little chance the molecules will achieve the necessary alignment, even at very cold temperatures. Sometimes no coherent crystalline structure forms until the temperature approaches −40° F. Below this temperature, all water molecules will freeze; indeed, gaseous molecules—vapor—will crystallize without passing through the liquid state at all, in a process that is known as sublimation.

Just as liquid droplets form around dust particles, so the droplets will freeze more readily if particular kinds of particles, called freezing nuclei, are present. Such particles, most notably the dust of certain clays, have a crystal structure similar to that of ice, and supercooled water molecules easily mimic their structure on contact and crystallize to form ice. Once some crystals have formed, water molecules diffuse onto the ice and the process of crystallization accelerates.

The ice crystal theory was accepted until the early 1940s as the explanation for virtually all rainfall. But as aviation expanded the knowledge of the upper atmosphere, it was discovered that clouds in the tropics are rarely cold enough, even in their upper reaches, to permit the formation of ice crystals. Some other mechanism had to be responsible for the water droplets' merging into raindrops, and scientists had to rethink the entire process in light of the new data.

Because water droplets of uniform size tend to be carried along by air currents at the same speed, collisions between them are rare. Even if they do collide, evenly matched droplets usually bounce off each other rather than combine to form a raindrop. Investigators decided that in the tropics some droplets must at the outset be larger than others. Scientists began to suspect that salt particles, wafted skyward from equatorial oceans by the tropical heat, might account for the larger droplets. (As anyone who has attempted to rattle a salt shaker on a humid day knows, salt attracts water vapor.) Experiments soon confirmed that large salt particles are aloft in the tropics and do indeed encourage the growth of correspondingly larger water droplets. Because larger droplets are heavier, they are the first to overcome the lifting provided by warm air; as they descend, the larger droplets collide and coalesce with the smaller, until a full-grown raindrop falls from the cloud.

Some scientists also believe that electricity can be a factor in raindrop coalescence. If two droplets have opposite charges—one positive, the other negative—they will combine more readily. Thus, the electricity in the cloud may well encourage the growth of droplets to raindrop size.

The story of precipitation does not end with the formation of a falling raindrop or snowflake; the intricate processes of the atmosphere have much more to do in determining the nature of the entity that finally reaches the surface of the earth. At times, precipitation is inhibited by the same updrafts that deliver the necessary moisture aloft in the first place. As snowflakes or raindrops fall, they pass through warm air that is still ascending. Although the water vapor in the rising air is cooling and condensing, the process of condensation itself releases what is known as latent heat—energy given off in the form of heat when matter changes its state from a gas to a liquid or from a liquid to a solid. Thus, while the lower pressure at higher altitudes is chilling the upward-moving air, the release of latent heat by condensation tends to keep the rising air warmer than the surrounding air and maintains the upward movement. Often, the air keeps rising until it enters a layer of warmer air in which the rising parcel no longer has buoyancy. In the absence of such a lid on convection, the upward movement of air within a cloud is always under way, retarding the fall of precipitation until the snow or raindrops grow large and heavy enough to overcome the updraft.

When a rain cloud moves across the sky, the first drops to fall toward earth pass through warmer, drier air below the cloud. This passage through

Rime ice forms as water droplets freeze onto stationary objects: A strong wind amassed a thick layer on the windward side of the lamppost above; the icy filigree on the fence at right is an accumulation of smaller droplets that were deposited by a light breeze.

the unsaturated air causes many of the small drops to evaporate again. The result on the ground is a random spattering of large raindrops. In a short while, the falling drops cool the ground and moisten the lower air, and a downpour begins. If the air near the surface is very hot and dry, as it is over a desert, all of the raindrops may evaporate before they hit the ground. When this happens, an observer on the ground may see what are known as virga, a curtain of rain or ice crystals that trails from a cloud only to vanish in midair.

Snowflakes, because of their feathery construction, are even more subject to the vagaries of the atmosphere. Some snowflakes fall at an extraordinarily slow pace; theoretically, a flake might take two days to reach the ground from an altitude of 10,000 feet (if it ever gets there at all). Quirks of temperature and humidity strongly influence the snowflake's journey earthward. If the air near the surface is dry, snow may reach the ground even when the temperature is above freezing; as some of it melts and quickly evaporates, the evaporation cools the lower air, and subsequent snowflakes traverse it unscathed. On the other hand, there are times when the saying "It's too cold to snow" holds true. Very cold air may not contain enough moisture to form any kind of precipitation. Frigid polar air is so incapable of holding water vapor that explorers exhaling a lungful of warm, moist air have seen it turn instantly into snow, which then falls gently on their boots.

Snow crystals that melt during their fall can cause the winter's most spectacular raiment, as well as its most miserable weather. When the melt-

An Avid Archivist of Snowflakes

In 1880, a young Vermonter named Wilson A. Bentley peered through a microscope and got his first glimpse of the delicate tracery of ice crystals that constitutes a snowflake. Bentley, then 15, was so fascinated with what he saw that he made a life's work of photographing and cataloguing the endless variety of shapes that snow crystals assume.

Although it was widely known that snowflakes form in crystalline patterns, Bentley was the first to record them on film, using techniques that were both ingenious and unique. Catching the falling flakes on a velvet-covered tray, Bentley quickly singled out promising crystals with a magnifying glass, then transferred them with a wooden probe onto glass microscope slides, smoothing them into place with the stroke of a feather. While still outdoors, or inside a chilly shed, Bentley placed each slide under a low-powered microscope linked to a camera and then photographed the magnified flake.

During 40 winters of painstaking work, "Snowflake" Bentley, as he came to be known, accumulated thousands of microphotographs—but he never found two snowflakes that were identical.

Standing outside his Jericho, Vermont, farmhouse, Wilson Bentley collects falling snowflakes during the winter of 1925. Arrayed around him is the apparatus he used to inspect and photograph snow crystals.

A few of the images published in Bentley's 1931 book *Snow Crystals* hint at the infinite variations on the basic six-sided pattern. Today, meteorologists recognize more than 80 categories of snow-crystal structure.

ed flakes pass through a deep layer of extremely cold air near the surface, the droplets refreeze; the result is sleet, that much-loathed winter hybrid. When the rain does not refreeze into sleet or ice pellets before it reaches the ground, the droplets may still be so cold that they freeze instantly upon contact with the frozen ground. If the sun happens to break through after this freezing rain, the glazed trees and bushes may sparkle brilliantly in a bejeweled wonderland, but the coating of ice is also dangerous; its weight can snap tree branches and power lines, and its slick sheen can cause havoc on the highways.

The endless variety of snowflakes is legendary, and although scientists insist that no law of nature forbids their duplication, two identical flakes have never been found. One search of epic proportions was conducted by a Vermont farmer and photographer named Wilson A. Bentley, who spent more than 40 years examining and photographing snowflakes through a microscope without ever finding two that were exactly alike. There are, however, recurring types of snow crystals that combine to form flakes. The crystals are broadly classified as either symmetrical or asymmetrical, and laboratory experiments indicate that their shape is determined by the temperature and moisture conditions of the cloud in which they form. If the cloud's temperature is above 27° F., the crystals probably will be hexagonal and flat; between 27° and 23° F., they are likely to be needle-shaped; and between 23° and 18° F., they will be hollow, prismatic columns. At lower temperatures they may be hexagonal, columnar or even fernlike.

Whatever their shape, snowflakes possess superb insulating qualities. As they descend, they act as acoustical baffles, damping sound vibration to produce the eerie quiet that is so characteristic of a gentle snowfall. Snow is also a thermal insulator: The flakes' ability to trap tiny pockets of air enables a blanket of snow to act much like a down comforter. During one cold snap in the Midwestern United States, the temperature on the snow's surface was −27° F., while just seven inches below the surface the reading was 24° F.—a change of 51° from the top to the bottom of the snow blanket.

The most formidable result of the various combinations of precipitation and turbulence is hail. The bane of farmers, and occasionally a threat to life itself, hailstones generally are a half inch in diameter, but can grow to the size of golf balls or even baseballs. The largest hailstone ever recorded, found on September 3, 1970, in Coffeyville, Kansas, measured 17½ inches in circumference and weighed 1.67 pounds.

In order for a hailstone to form, ice crystals must remain airborne in a cloud of supercooled water droplets long enough to grow to thousands of times their normal size. Then they must fall rapidly through the air without melting. Usually, the only clouds turbulent enough to produce hail are thunderheads, which form principally in summer, when surface heat generates the strongest updrafts. These magnificent cumulonimbus giants often soar to the beginning of the stratosphere, where temperatures from −58° to −112° F. create an ice-laden cap. There are two theories about hail formation in such a cloud. The older and more widely known view is that ice pellets form at the top of a cloud, fall almost to its base and there acquire a layer of supercooled water. Then they are caught in an updraft, and as they rise the water freezes, enlarging the ice pellets. According to this theory, hailstones result from several such round trips, each one adding a new layer of ice until

The surprisingly short life of a cumulus cloud begins when the water vapor in a rising parcel of warm air cools and condenses. The cloud soon drifts away from the source of warm air and evaporates, often in as short a time as 15 minutes, to be replaced by a new cloud.

WIND

RISING AIR

RISING AIR

Fair-weather cumulus clouds populate the sky over the Colorado River in southwestern Arizona. The clouds appear over rising columns of air heated by patches of sun-warmed earth.

the hail plummets to earth. Many large hailstones have an onion-like structure, which supports this view of repeated circulation.

The second theory, developed in more recent years, holds that the hailstones make only one descent, but that the fall is very slow. The updraft supports the hailstone nuclei for a considerable time—from 10 to 20 minutes—while they are buffeted by other snow crystals and supercooled droplets. Eventually, the incipient hailstones absorb so much heavy ice that they overcome the updraft.

It is likely that both models of hail formation are correct; at times both processes may even be at work in the same storm. "The atmosphere," one meteorologist remarked, "is big enough to accommodate both of them."

While the most daunting challenge confronting modern students of cloud physics has been to explain what happens in nimbus clouds, recent discoveries have illuminated countless details about the other categories Luke Howard identified almost two centuries ago—cumulus, stratus and cirrus clouds. A cumulus cloud is now understood to be the result of vertical mixing, a process that begins when a parcel of air is warmed by

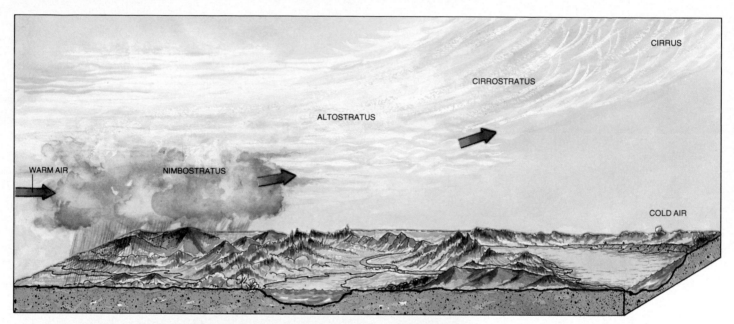

CIRRUS

CIRROSTRATUS

ALTOSTRATUS

WARM AIR

NIMBOSTRATUS

COLD AIR

The basic types of layered clouds often appear along a warm front in the sequence diagramed above. They range from high, wispy cirrus clouds to the gloomy, ground-hugging nimbostratus rain clouds pictured at left.

sun-baked fields or even large parking lots. The invisible parcel of warm, rising air, called a thermal, cools until its temperature reaches the dew point and its water vapor condenses, suddenly becoming visible as a cloud. Because thermals occurring in the same region cool at the same rate, cumulus clouds tend to form at the same altitude and their bases appear flat. Their vertical extent, however, depends on the strength of the updraft and the ·temperature structure of the troposphere. The largest of them, the stormy cumulonimbus clouds, typically climb thousands of feet into the sky.

A large cumulus cloud is composed of many smaller air parcels, each forming a lobe on the cauliflower-shaped mass. Each lobe, or "floret," has a life span of only 10 to 15 minutes. Powerful updrafts carry the water droplets through the center of the cumulus cloud to the top, where (unless the cloud is high enough to reach the frigid realms where the droplets crystallize) they evaporate. A portion of the air then flows outward and sinks again, creating downdrafts around the border of the cloud. Meanwhile, a steady horizontal wind often pushes the entire assemblage sideways so that the cloud is transported away from the rising thermal. Without an updraft to feed it, the cloud breaks up and vanishes while another one comes into being behind it, over the thermal.

Entire cumulus systems often go through a daily cycle. Created by sun-induced convection, cumulus clouds typically begin to appear in midmorning. By noon, legions of small clouds may cover the sky, and by late afternoon they may be numerous enough to block much of the sunshine. During the afternoon, especially in the summer, a few may develop into the great cumulonimbus towers that spawn thunderstorms. Sundown brings an end to the solar heating that powers convection, and the clouds dissipate.

For the most part, these cumulus life cycles occur over land, because the ocean generally warms too slowly and too evenly to foster the heat concentrations needed to form distinct updrafts. Every day, air travelers crossing a coastline can observe a land-based cloud pattern that ends fairly abruptly at the beach. Cumulus clouds do appear from time to time in midocean, however, because of the lifting associated with low-pressure systems, with warm ocean currents or with thermals over islands. Near the Equator, where the trade winds converge and moisture-laden air rises quickly, great thunderstorms and severe downpours occur frequently.

The gray cloaks of stratus cloud that layer the sky from horizon to horizon form when the air is too stable to permit the growth of cumulus clouds. While cumulus clouds result from warm updrafts rising through a colder environment, stratus usually appear when a warm air mass collides with a colder one. The lighter, warm air moves upward and over the dense, cold air near the ground, cooling until its moisture condenses into a vast and leaden cloud. The warmer air aloft prevents the cold air from rising; indeed, by definition the term "stratus" implies a complete lack of vertical air movement. Normally, low stratus clouds will merely blanket the sky, without producing rain; but when the clouds are high enough or thick enough, ice crystals will form and begin to produce precipitation.

To further distinguish the different types of stratus clouds, some additions to Luke Howard's nomenclature have proved useful. Nimbostratus are rain-producing clouds found within 5,000 feet of the ground. Altostratus, meaning "middle layer," typically form at altitudes of between 10,000 and

20,000 feet, and cirrostratus clouds are usually found in a thin wispy layer above 20,000 feet.

The very lowest stratus clouds have a far simpler designation—fog. Because of its proximity to the human realm, fog is in some ways the most dangerous of all clouds. Throughout history, hapless mariners blown near a rockbound coast have read their own death sentences in the silent mist that enveloped their sailing ships. Modern electronic instruments now enable most captains to navigate safely through a blinding fog, but extreme caution is still the watchword on the ship's bridge when a fog sets in.

In an earlier time, an airplane pilot circling a fog-shrouded airport might measure the remaining moments of his life by the needle on his fuel gauge. In the Allegheny Mountains of north-central Pennsylvania, where successive parallel ridges trap cool, moist air, pilots flying the mail in the early 1920s frequently came to grief amid both fogs and low-lying clouds that earned the sardonic designation "cumulogranite." In a single 22-month period, an area that pilots came to call the Hell Stretch—located between Bellefonte, Pennsylvania, and Cleveland, Ohio—claimed 26 lives. Fog remains a serious hazard even to modern jetliners equipped with sophisticated electronic equipment. In 1977, fog was a major factor in the collision of two 747 jets on an airfield in the Canary Islands—the worst disaster in airline history.

The condensation of water vapor into fog is often caused when moist air is

The early-morning sun reveals a carpet of valley fog near a village in the Austrian Alps. Cool, moist air settled to the valley floor during the night and cooled further until its water vapor condensed into fog.

cooled by direct contact with chilly ground. The earth cools rapidly on a night when no clouds are present aloft to trap or slow the radiation of its heat into space. In such conditions the air near the ground can quickly reach its dew point: The result is a radiation fog, so called because it is caused by the loss of the earth's heat through radiation.

A slight breeze aids the formation of fog; in absolute calm, fewer air molecules come into contact with the cool surface, and a stiff wind may disperse the fog entirely, but a breeze that mixes the damp air gently may stir the fog to a height of several hundred feet. Radiation fog tends to occur in low-lying areas and in enclosed valleys—like those of the Hell Stretch—where cold air collects readily. It is usually thickest just after dawn, when the initial heat from the sun creates a gentle convection that circulates warmer moist air into the colder air near the ground. Then, within a few hours, the light of the new day will generally evaporate the fog.

A different kind of fog forms over a lake or a quiet river when cold air drifts over warmer water and is both warmed and moistened with water vapor. This warmed air mixes with the slightly colder air above it, and when the mixture reaches its dew point, a steam fog forms slightly above the surface of the water. Such fog appears when the water is roughly 5° to 10° F. warmer than the air, which is frequently the case in the early autumn. Steam fog is known in polar regions as arctic sea smoke, because it occurs over cracks in the icecap. The water beneath the ice, though near

Steam fog wafts across Penobscot Bay in Maine. Also known as arctic sea smoke, this particular type of fog appears over open water in cold weather, as frigid air causes vapor rising from the warm water to condense.

freezing, is warmer than the air overhead, and the rising vapors condense almost immediately.

Fogs created by radiation and evaporation are generally local phenomena, but another well-known kind of fog may cover large areas. Called advection fog, this soupy mix occurs when a humid air mass moves across a cold surface. In the Atlantic Ocean off the coast of Newfoundland, air warmed by the Gulf Stream frequently flows toward the northwest and comes into contact with the frigid Labrador Current; the resulting fogs near the Grand Banks are among the most persistent in the world. Off the coast of California, warm, moist air from the open ocean drifts over the cold water of the California Current, creating the slow-motion wave of fog that regularly pours over the coastal hills to engulf San Francisco. Although some residents find this fog tiresome, the moisture is vital to northern California's famous redwood trees.

Inland, clouds known as upslope fogs form when a low moving blanket of warm air is gradually elevated toward a cooler altitude by the upward slope of the land. Upslope fogs occur frequently on the sides of mountains: New Hampshire's Mount Washington is socked in for at least 300 days every year. On America's Great Plains, a peculiar kind of upslope fog develops when a southeast wind blows moist air from the Gulf of Mexico toward the Rockies. A vast blanket of fog may form over western Kansas or Nebraska as the gradual upward tilt of the land west of the Mississippi River conducts the air to an altitude where the air is cool enough to cause condensation.

The highest clouds described by Howard were cirrus clouds. Rarely seen below 20,000 feet, they are composed entirely of ice crystals. Their curled look is occasioned by crystals that fall from the main cloud and trail behind it in the calmer wind below. Wispy cirrus clouds can be produced by small-scale vertical motion—or they can be the remnants of the top of a thunderstorm cloud. A variation—cirrostratus clouds—forms when large sheets of air ascend slowly into the upper troposphere. Cirrocumulus clouds are another product of convection currents; they have a rippled appearance and produce what meteorologists call a mackerel sky. Cirrus clouds usually signal the presence of powerful updrafts in the distance, and for this reason they are reliable heralds of unsettled weather conditions.

Cirrus clouds can also be man-made. When a jet airplane flies at high altitude through air that is particularly cold, it often leaves a trail of condensation called a contrail. Normally, the clear, cold air through which the jet passes is too dry to permit cloud formation. But the jet engine spews out water vapor, which freezes in a matter of seconds, forming a trail of ice crystals behind the plane. Just like its natural cirrus cousin, a contrail provides a clue to the next day's weather. A contrail will vanish rapidly if the air is relatively dry, as it is in a high-pressure zone, but if the air aloft is moist, the contrail may linger for an hour or more. A contrail that vanishes swiftly is thus an indicator of fair weather, while a persisting one suggests moist air, and perhaps an incoming low-pressure air mass—the storm maker.

Modern students of the atmosphere have, of course, identified and explained clouds and similar atmospheric conditions not dealt with by Luke Howard. One strange apparition that he may never have seen is the lenticu-

Carried along by strong winds, gauzy cirrus clouds veil the sky above rock formations in Utah. As some of the ice crystals that make up these high-altitude clouds drift into the calmer air below and trail behind, they form long, horizontal streamers.

lar, or lens-shaped, cloud that appears above or on the downwind side of a mountain. This cloud reveals the crest of a wavelike, invisible air current that has been deflected upward by the mountain. As the air moves into the cooler realms near the top of the mountain, the water vapor condenses and the lens-shaped cloud forms. When the air sinks again, as it will to follow the contour of the land, the droplets evaporate. Yet the cloud itself stays in place because the wind continues to flow over the mountain, and as fast as the downwind edge of the cloud evaporates, water vapor condenses to create new cloud upwind.

Many solitary lenticular clouds have spawned reports of flying saucers. When smaller lenticular clouds form downwind of the main cloud as the air rebounds upward, an imaginative observer might perceive a squadron of alien spacecraft. In fact, the air has merely risen again to the condensation level, and a diminishing series of lesser air wavelets has brought into being a small chain of lenticular clouds.

Haze and dew, two other familiar atmospheric conditions, are created when the recipe for cloud formation is not quite complete. Haze dims the sky when an abundance of airborne particles scatters the sunlight. If the air's moisture content is too low or if the temperature is too high, little or no vapor will condense around these aerosols to form a cloud. Without a wind to blow the particles away, the haze persists, its color an indication of the nature and size of the particles. A blue haze is often caused by a chemical combination of atmospheric ozone and terpenes—hydrocarbons thrown off by vegetation. The Blue Ridge Mountains of the Eastern United States probably owe their name to the tint of this kind of haze. A gray tinge may come from larger particles of soil, salt or other minerals. At the shore, the breaking waves and crashing surf throw up clouds of salt spray, and billions of salt particles are left in the air when the water in the spray evaporates. The resulting haze may extend miles out to sea.

Like fog, dew forms when water vapor is cooled by a cold surface and condenses. The droplets are confined to grass, leaves or other objects near the ground when there is no vertical mixing to carry the moisture upward or when the moisture content of the air is too low to expand the condensation process into a full-fledged fog. At times, vegetation also contributes to the dew through a process called transpiration—the evaporation of moisture from plant cells.

Together, the two processes of condensation and transpiration can pro-

A lenticular cloud caps a peak in the Pacific Northwest. As air—here moving from right to left—is forced upward to clear the summit, it cools and forms the cloud. The air sinks and warms on the other side of the mountain, causing the cloud to thin out and finally vanish.

duce a foot-soaking film of water even on a well-trimmed lawn, but this pearly sheen can also create a vision of surpassing delicacy and beauty—a dew halo. Throughout the first few hours of the morning, before the dew evaporates, a person standing with his back to the sun may see a distinct halo surrounding the head of his shadow. Upon witnessing this strange effect for the first time, one 16th Century sculptor and writer interpreted it as a sign of divine recognition of his genius. However, scientists now believe that the process is less personal and that the halo is caused by the bending and reflection of sunlight by the tiny dewdrops. The sunlight enters the front of the drops, bends very slightly and then is reflected from the rear of the drops toward the observer. A similar phenomenon accounts for one of the most spectacular of all interplays between light and moisture—the rainbow.

These great arcs of color, created when sunlight strikes a rain shower, have been the subject of speculation for thousands of years, and their fundamental cause was worked out long before cloud dynamics were fully understood. As early as 1304, a German Dominican monk named Theodoric came up with two inspired deductions about rainbows. First, he had the imagination to see that individual raindrops provide the key to the puzzle. Second, he understood that both reflection and refraction of light are needed to create a rainbow. Using a water-filled glass globe to simulate a giant raindrop, Theodoric observed that while most light passed right through the globe, some of the incoming light was bent by the convex outer surface, reflected by the inner concave surface and then bent again as it exited.

Further scrutiny of his giant raindrop revealed that, at any one viewing position, only one color was visible; if the observer moved slightly, other colors would appear. From this Theodoric concluded that each color visible in a rainbow comes from a different set of drops; it is the totality of millions of drops, each sending out a particular color to the viewer, that creates the visible rainbow. Thus, no two people ever see exactly the same rainbow: Each views a different set of drops at a slightly different angle.

Theodoric next went on to explain the appearance of the secondary rain-

Morning dew beads a flower and a sluggish fly. A clear sky permitted the fly and its perch to cool sharply overnight; the chill condensed moisture from the air that surrounded them.

Born of the combination of sunlight and rain following a summer thunderstorm, a rainbow spans the cactus desert near Tucson, Arizona. A faint secondary bow appears just above the main rainbow, its colors in reverse order.

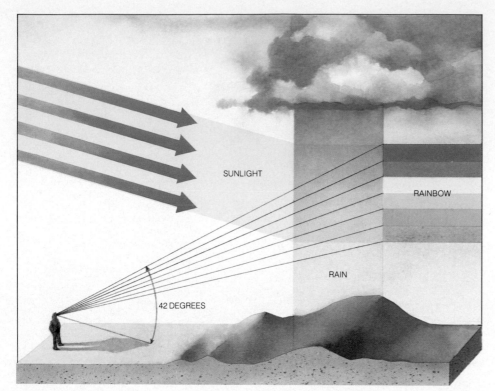

Labels in image: SUNLIGHT, RAINBOW, RAIN, 42 DEGREES

bow that sometimes appears above the main one. This, he said, is caused by light entering near the bottom of the droplet and reflecting twice off the back inner surface, much as a bullet might ricochet twice inside a steel bowl before emerging again. The secondary rainbow is not as bright as the first because some of the light is lost in the second reflection. The secondary reflection also causes the color sequence to reverse, just as a mirror reverses an image: Red is always at the top, or outside, of the primary rainbow, but at the bottom, or inside, of the secondary bow.

Theodoric's findings remained largely unnoticed until the 1600s, when René Descartes began his own investigation of rainbows. Descartes set out to discover why rainbows were visible only at an angle of 42 degrees from the path of sunlight. If an observer looks at the top of his shadow, then moves his eyes vertically through an angle of 42 degrees, that is where he will see the rainbow. After experimenting with a water-filled globe similar to the one Theodoric had used, Descartes determined mathematically that when two refractions and one reflection are involved, the 42-degree angle was the path along which the least scattering, or maximum focus, of the light occurred.

Sir Isaac Newton discovered the reason for the color spectrum of a rainbow while he was experimenting with glass prisms in 1666. Newton showed that each color in the spectrum has its own optimal angle of refraction. When sunlight passes through a raindrop, refraction does not weaken the light but simply sorts it out by wavelength, sending each color along a slightly different path. To underscore this, Newton went on to perform a historic experiment: He used a second prism to recapture the divided colors and merge them back into white light—thus proving for the first time that white light is a composite of all the other colors.

In time, scientists also discovered that the size of the raindrops helps to determine the intensity of the colors. When drops are especially small, the colors tend to overlap in such a way that orange and pink tones are at the edges instead of reds and violets. In a fog, the droplets are so small that the colors merge completely and the resulting fogbow is white.

Every rainbow is a segment of an imaginary circle whose center is indicated by the shadow of the observer's head. Therefore, no one on the ground ever sees a completely circular rainbow, or sees a rainbow segment unless the sun is low enough to provide the necessary angle between the shadow and the horizon. Circular rainbows can be seen from aircraft, however, and small, nearly circular spectra can be created with the spray from a garden hose.

Just as the investigators of clouds and rain had found themselves ever more enmeshed in complications as they pursued the secrets of everyday occurrences, so did the students of rainbows find their inquiries to be perpetually unfinished. Having probed the optics of common rainbows, for instance, they had to deal with the further intricacies of supernumerary arcs. These very faint bands—as many as four—are sometimes visible on the inside of the primary bow. They are the result of the complex interaction of light waves that travel in parallel paths from the area of the rainbow to the viewer's eye. When the waves are out of phase—that is, when the crest of one wave coincides with the trough of the one next to it—they tend to cancel each other out, darkening and often smudging their colors. When in phase, and the crests and troughs match, color intensity is enhanced.

However complex rainbows may seem when explained scientifically, a detailed understanding of how they are created ultimately magnifies their legendary glory. A lustrous rainbow against a backdrop of steel blue storm clouds is inevitably breathtaking, but when enhanced by knowledge, it comes to signify much more: the grand rhythms, the countless variations and the minute complexities of the global web of relationships among water, light and air that encloses and supports the earth and all life on it. Scientists have thus accomplished something that Shakespeare thought to be a useless endeavor: "To add another hue unto the rainbow." **Ω**

A GALLERY OF CLOUDSCAPES

The stuff of all clouds, tiny water droplets or ice crystals adrift in the atmosphere, appears when air is cooled past
its dew point, the temperature at which
its load of water vapor condenses. The
updrafts on a sunny day produce small,
transitory puffs of cumulus cloud; the
vast forcing of air upward along a frontal
system can cause stratus clouds to form
in layers hundreds of miles long.

Countless movements of the air can
further sculpt these basic kinds of cloud
into formations of breathtaking beauty
and variety. When a layer of rapidly
moving air overlies a calmer region, any
clouds that form are whipped into billowing waves or long streamers. Air
coursing over mountainous terrain can
shape lenslike clouds of startling sym-
metry. When temperature differences
develop between the top and the bottom
of a shallow layer of cloud, many small
convection currents may stir the layer
into a pattern of small, puffy cells.

Even the air movements that hasten a
cloud's evaporation can lend it a special
glory in decline. When the updrafts that
nourished a thunderstorm ebb, air currents sometimes mix dry air into the
body of a cloud, causing it to lose its
moisture and reducing it to tatters of
cloud that trace the outlines of the ex-
tinct thunderhead.

A thin layer of altocumulus cloud displays
a woolly pattern produced by convection. Each
of the tufts is a circulation cell in which
warm air at the base of the cloud rises to the top
of the cloud, cools and sinks again.

Cumulus clouds shed trails of ice crystals, known as virga, that evaporate without melting long before they reach the ground.

Puffy cells of cloud in an altocumulus layer owe their thickness to strong convection currents in a deep layer of moist air.

Bands of cirrocumulus clouds resembling rows of fish scales form a mackerel sky. Swift winds above the cloud layer have rippled it in much the same way as a breeze ripples water.

Mammatus clouds bulge under the high
anvil of a cumulonimbus cloud. Such formations
are produced in severe thunderstorms by
downdrafts of dense air cooled by the
evaporation of countless cloud droplets.

Holes gape in a thin layer of stratocumulus
cloud as supercooled droplets freeze into wisps of
ice crystals. Vibration from a passing plane
may have triggered this striking transformation.

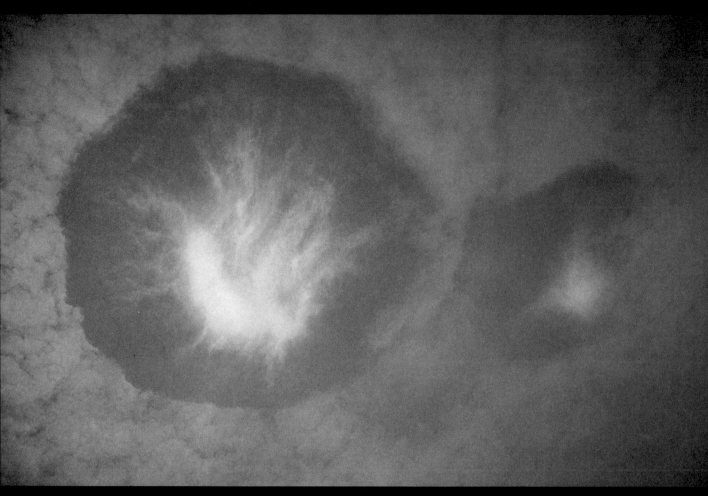

Gossamer trails of cirrus cloud called mare's-
tails form when ice crystals fall from fast-moving
clouds into calm air below and are left behind.

Stratocumulus clouds (*right*) billow across
an early-morning sky. Such patterns, shaped by
the wavelike motion of a layer of moist air
that has passed over an area of rough terrain, can
persist for several miles downwind.

Scattered altocumulus clouds dapple the sky over
the Alaskan tundra an hour or two after an
afternoon thunderstorm. These remnants of the
thunderhead will soon vanish altogether.

Chapter 5

"A FULIGINOUS AND FILTHY VAPOR"

In a sense, the atmosphere has always been polluted. It is, after all, a complex mixture of gases and particles, some beneficent and some noxious. Long before fire was discovered and humanity began to contribute to the problem, the air was replete with soil dust, oceanic salt particles, pollen, bacteria, smoke and gases from forest fires, volcanic debris and innumerable other substances. And to this day, most airborne materials still come from these natural sources. "A great deal of poetic nonsense has been written about country and mountain air," research chemist and author Donald Carr once noted. He described the air of verdant mountains and hills as "a very complex perfumed mess." Indeed, organic vapors rising from upland slopes in the Southeastern United States earned the mountains there a memorable nickname, the Great Smokies. Similarly, in the Middle East, dust blown aloft from the desert produces so much haze that the reflection of the tinted sky caused a body of water there to be called the Red Sea. When compared with the volume of natural emanations, the pollution generated by modern civilization seems insignificant.

Yet its effects are staggering. Increasing amounts of contaminants entering the air from man-made sources have been found in recent years to be doing far more than occasionally fouling the air and choking the residents of one locality or another: They are causing ancient statuary on the Acropolis to crumble, Egypt's pyramids to dissolve and the fish in remote, seemingly pristine mountain lakes to die. The more the threat has spread—across countries, continents and indeed the very planet—and the more scientists have learned about it, the more complex the problem has proved to be. That the atmosphere and most of the life it shelters has survived thus far is largely due to the size of the earth's gaseous envelope and its ability to neutralize harmful substances and cleanse itself from time to time with rain.

The most devastating pollution crises of modern times have been brought about by inversions, atmospheric conditions that stifle natural dispersion. The basic ingredient of an inversion is the presence of a warm layer of air aloft that prevents the surface air—and the pollution it contains—from rising. One type of inversion occurs when a locale normally subject to frequent fogs—London, for example, or Pittsburgh—comes under the spell of a large, slow-moving high-pressure air mass that sweeps warm air in at high altitude; the mass can trap the air beneath it for days. An inversion can also be caused by a geographical quirk. Surface air that is cooled by a nearby body of water may be trapped beneath a warm air mass if both layers of air

Triggered by man-made heat, a cumulus cloud forms above the smokestacks of a western Pennsylvania power plant in a graphic demonstration of the impact that industrial activity has on the atmosphere.

are confined within an encircling range of hills or mountains. Such is the lot of Los Angeles and a number of other coastal cities, including Capetown, South Africa, and Santiago, Chile. This type of inversion normally creates moderate, pleasant temperatures in regions that would otherwise be much warmer. However, it also invites severe pollution build-up.

The forms of man-made pollutants most often trapped by inversions are gases and tiny airborne particles or droplets referred to by scientists as aerosols. Although most aerosols are microscopic in size, more than a half billion metric tons of them enter the atmosphere each year as a result of the incomplete combustion of fossil fuels. Carbon, ash, mercury and lead particles escape into the air whenever the burning process is inefficient—and virtually all conventional combustion processes are inefficient to some degree. Most aerosols are found in the lowest two miles of the atmosphere; they become especially concentrated when they are trapped under an inversion. Extremely large particles, such as coal ash, may remain in the air only a few minutes, but others, including lead and smoke particles, can stay aloft for weeks. Sometimes rain brings them down, sometimes gravity alone, but most of the windborne particles are deposited by impact when they collide with trees, buildings, hills and other topographical obstacles. For this reason, it is common to find almost as much dust on the underside of a leaf as on the top.

Not all aerosols are hazardous to humans; the largest are likely to be screened out by very efficient natural filters in the nose and throat, and the smallest simply pass in and out of the lungs without doing any damage. It is those in the middle range, such as sulfates, nitrates and soot, that are most dangerous, because they often lodge in the minute air passages of the lungs. There they may aggravate bronchial disorders and possibly generate serious illnesses.

Even when most aerosols are consumed in the burning process, polluting gases remain. In fact, the volume of man-made gases entering the atmosphere every year is almost 10 times that of aerosols. The burning of fossil fuels produces a variety of potentially harmful gases, including carbon dioxide, sulfur dioxide, carbon monoxide and nitrogen oxide.

The most abundant of these gases is carbon dioxide, an ordinarily innocuous compound vital to all plant life. Natural production of carbon dioxide far outweighs industrial emissions (every breathing creature exhales carbon dioxide), but the addition of industrial carbon dioxide to the natural supply may have a profound, long-term effect on global climate. This ubiquitous gas plays an important role in regulating the earth's temperature. In a process called the greenhouse effect, it tends to reflect back toward the earth heat that would otherwise escape into space. Thus, too much carbon dioxide could cause a worldwide warming trend.

If carbon dioxide poses perhaps the greatest future threat among the pollutant gases, sulfur dioxide has been the most troublesome during the last several centuries. Although sulfur dioxide in small amounts is a harmless natural component of the atmosphere—volcanoes produce it sporadically—it has deadly potential in the high concentrations spewed out by large-scale burning of oil and coal. (The combustion of 10 tons of bituminous coal has been known to produce a ton of sulfur dioxide.) When sulfur dioxide mixes with water—in fog or rain, for example—sulfuric acid forms; this extremely corrosive and irritating liquid can eat away steel,

The photomicrographs at right portray various air pollutants, enlarged 30 to 100 times. Some specimens were filtered from the air; others were collected from the surfaces on which they accumulated. For clarity, the colors have been enhanced with a polarizing filter.

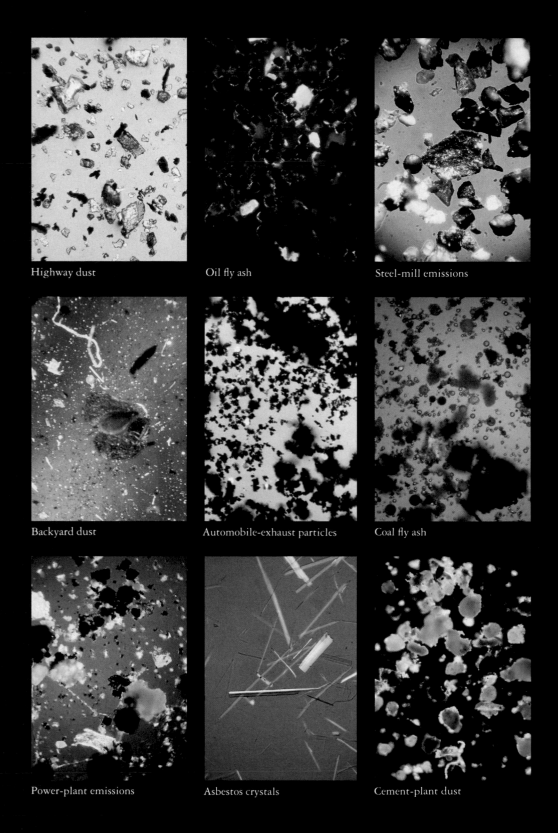

Highway dust

Oil fly ash

Steel-mill emissions

Backyard dust

Automobile-exhaust particles

Coal fly ash

Power-plant emissions

Asbestos crystals

Cement-plant dust

Ash-laden clouds billow skyward during a 1974 eruption of Guatemala's Fuego volcano. Like many other volcanoes, Fuego was a heavy polluter; suspended particles crimsoned Northern Hemisphere sunsets for months.

dissolve limestone and marble, and work many forms of mayhem on delicate human tissues.

In 12th Century England, deforestation led to the use of coal in manufacturing. Most royal subjects could do little about the resulting unpleasant smell, but in 1257, Queen Eleanor, wife of King Henry III, refused to stay in Nottingham Castle because of the choking air sent up from the coal fires in the town below. For the most part, however, economic prosperity proved more popular than clean, cold air, and coal fumes became a pervasive part of England's atmosphere for the next 700 years.

Londoners especially relied on home coal fires in the hearth to dispel the chill of the city's fogs, which were in part caused by particulates emitted from chimneys. In 1661, the prominent Londoner John Evelyn published a pamphlet entitled *"Fumifugium,* or the Inconvenience of the Aer, and Smoke of London Dissapated; together with Some Remedies Humbly Proposed." He addressed his tract to King Charles II.

"Sir," he began, "it was one day as I was walking in Your Majesty's

SOURCES OF PARTICLES IN THE ATMOSPHERE

NATURAL SOURCE	Millions of tons per year
Sea Salt	1,000
Soil Dust	200
Volcanic Eruptions	4
Forest Fires	3
TOTAL	**1,207**

MAN-MADE SOURCE	Millions of tons per year
Combustion of Coal	36
Combustion of Oil	2
Combustion of Wood	8
Incineration	4
Agriculture	10
Cement Manufacture	7
Iron and Steel Manufacture	9
Miscellaneous	16
TOTAL	**92**

Although far fewer particles are discharged into the atmosphere by human activity than by natural processes, the man-made particles are often finer in size, slower to settle out of the air and more dangerous to health.

Palace at Whitehall (where I have sometimes the honor to refresh myself with the sight of Your Illustrious Presence, which is the joy of Your People's hearts) that a presumptuous smoke issuing from near Northumberlandhouse, and not far from Scotland-yard, did so invade the Court, that all the rooms, galleries and places about it, were filled and infested with it, and that to such a degree as men could hardly discern one another from the cloud, and none could support, without manifest inconvenience." Most Londoners, Evelyn added, "breathe nothing but an impure and thick mist, accompanied by a fuliginous and filthy vapor, corrupting the lungs, so that catarrhs, coughs and consumptions rage more in this one city, than in the whole Earth." With this delicate plea he hoped to persuade the King to have factories moved out of the city and aromatic trees and shrubs planted in London to reduce the stench of home coal fires.

Charles II took no action. Coal use increased, and thereafter sporadic fogs silently claimed scores of lives. Occasionally, officials gathered to ponder the problem, but even when laws were passed regulating industrial emissions, no one paid the least bit of attention. England prospered during the next three centuries, and later generations learned to shrug off air problems with soothing aphorisms. "Muck is money," they said.

The Industrial Revolution brought great wealth to England and Europe in the 18th and 19th Centuries, but with material progress came a stupendous increase in air pollution—and respiratory problems—because most factories burned coal. Many doctors understood the direct link between air pollution and a variety of serious illnesses. Nevertheless, the true magnitude of the threat was not appreciated until the third decade of the 20th Century.

On December 1, 1930, a thick fog settled over the heavily industrialized Meuse Valley in Belgium. Trapped by an inversion in the steep-sided valley, the soupy mix lingered for four days while smoke from steel mills, glass factories, lime furnaces, power stations and fertilizer plants transformed the fog into a suffocating blanket of noxious fumes. By December 3, thousands of people were choking, vomiting and gasping for breath, and by the time the smoggy mix lifted on the 5th, hundreds were seriously ill and 60 had died. A subsequent investigation revealed that the smog had probably held more than 30 impurities; the most damaging were sulfur dioxide and sulfur trioxide gases, along with hydrofluoric acid—a by-product of iron-ore smelters.

The penalties for fouling the air seemed to come in rapid succession in the middle of the 20th Century. Eighteen years after the Meuse Valley incident a similar disaster occurred in Donora, Pennsylvania, a heavy-industry town some 30 miles south of Pittsburgh. Nestled along a bend of the Monongahela River amid low rolling hills, Donora was the site of a steel mill, a zinc-reduction factory and a sulfuric-acid plant—all gross polluters. An inversion that developed in late October 1948 trapped the pollutants for three days. By the end of the third day, when rain finally washed the smog from the sky, almost 6,000 people were ill and 20 had died.

It was the first time that a smog infestation had struck a large city with such shocking effect. Just four years later, long-suffering Londoners had to account in full for their profligate coal-burning ways.

Wednesday, December 3, 1952, was a lovely winter day. Fleecy cumulus clouds filled the sky, and a fresh wind off the North Sea steadily cleared away the city's smoke. Londoners strolled in the parks and basked in the sunshine. The weather was the result of a huge high-pressure area centered over the western part of the country.

By Thursday morning, the anticyclone had moved closer to the city, the temperature had dropped and the air was distinctly moist. The diminishing wind had shifted to the north-northwest, and high clouds had formed at about 10,000 feet, a sure sign of warm air moving in aloft. An inversion was in the making. Here and there pockets of fog appeared, particularly over the Thames. The city began to reek of smoke as soot, gases and specks of ash from thousands of chimneys began to collect in the air. The larger particles began dropping on rooftops and in the streets.

On Friday, residents awoke to find the city enveloped in a dense yellow smog. To ward off the morning chill, Londoners did what they had always done; they lighted new coal fires in their grates and turned on electric stoves and heaters, causing coal-burning power plants to pour still more smoke into the thickening air. The wind died and the smog layer hugged the ground; at elevations of only a few hundred feet the air was clear, but at street level, visibility was limited to a few feet. Buses could proceed only when the conductor walked in front carrying a flare to guide the driver. One bicyclist heard a strange honking sound and prudently moved over to the side of the road. Suddenly, a swan materialized out of the mists, squawking its distress as it searched in vain for the Thames, which was about a half mile distant. As the day wore on, people began to notice a burning in their throats and a tightening in their chests. Those attempting to ride a bicycle through the muck found themselves covered with soot. The smog was even beginning to seep into homes, offices and public buildings.

To some residents, it was almost a lark, the sort of crisis Londoners had learned to endure cheerfully. But throughout the day, the smog grew darker, and by dawn on Saturday it was an ugly dark brown pall. By this time, the ravages of the sulfuric pollutants were widespread. Hospital emergency rooms were jammed with those suffering from respiratory or cardiac distress. Elderly people, their frail systems unable to endure the foul atmosphere, quietly perished in their rooms. Even the young were falling ill.

Sunday, the third full day of the attack, was just as bad as Saturday. Housewives found that the smog invading their homes left a black, sticky film on everything it touched. London's airports had shut down; barge traffic on the Thames was at a standstill; and in the streets, few vehicles moved. In many neighborhoods, visibility was less than one foot. The death toll, already into the hundreds, was mounting rapidly. Some of the stricken were turned away from overcrowded hospitals, and others died on the way as ambulances crept slowly through the miasma. One of the city's morgues accepted so many bodies that it ran out of shrouds. Doctors making house calls found themselves hard-pressed to keep up with the requests for aid—if they were able to reach their patients at all. One physician who kept getting lost in the gloom suddenly got the idea of asking a patient who was blind to guide him. The blind man did so unerringly. "All right, careful now, Doctor," he would cry out, "just a short step down here." Curiously, despite the obvious scope of the disaster, no official warnings or instructions were issued in the press or on the radio.

In the feeble daylight of Monday morning, December 8, the first sign of a break appeared: The air began to stir as a low-pressure center approached.

Holding a flare, a bus inspector guides a driver through the great London smog of December 1952. An inversion trapped fog and industrial pollutants over the city for four days.

Still the fog persisted. Movie-goers, hoping to escape the smog effects that evening, found that they could not even see the screen, and an opera performance had to be called off because the singers could not see the conductor.

Finally, on Tuesday—after four horrendous days—the smog dispersed and London began to count the casualties. Some 4,000 people had died, most of them elderly, and many thousands more were suffering from new or aggravated respiratory ailments. Shocked, Londoners at last resolved to bring an end to the air-pollution peril that had for so long literally hung over their heads. There was no way they could prevent inversions, but they could curb the city's outpourings of pollutants. Soon Parliament passed legislation establishing stringent new standards for smoke emissions, and provided government aid for the conversion of homes and industries to the use of low-sulfur fuels such as fuel oil and natural gas.

London experienced another inversion-related smog attack in 1956. From a meteorological standpoint, this inversion was even worse than that of 1952, but casualties were relatively light because the new regulations were working. A decade later, London was a city transformed; some said that it was fully 80 per cent cleaner. Wash hung out to dry now stayed clean, newly scrubbed public buildings remained free of soot, and the incidence of death from respiratory ailments plummeted.

Similar air-pollution restrictions were adopted in many industrial towns in the United States, and the results were equally encouraging. Most dramatic were the strict new rules put into effect in the "Smoky City" of Pittsburgh, just downriver from Donora. Realizing that the future of their community was imperiled by the noxious air, citizens implemented an urban renewal program in the late 1940s and early 1950s. The core of the program was a set of tough smoke-abatement provisions. Within a few years, Pittsburgh's atmosphere was greatly improved.

It now seemed that air pollution could be controlled easily by sensible laws. Actually, only one symptom of this great modern ailment had been relieved, and that only temporarily.

The true scope of the air-pollution predicament was about to be glimpsed in Los Angeles, where conditions were almost perfect for a dramatic build-up of a new generation of airborne contaminants. The weather of the Los Angeles area is dominated by the eastern edge of a high-pressure system that tends to hover over the Pacific Ocean between Hawaii and California. Some 2,000 miles in diameter, the air mass continually wheels warm, equatorial air toward the United States. As it approaches the coast, the lower air—usually within 500 to 1,000 feet of the surface—is cooled by the water beneath it. Thus, the air sweeping in off the ocean is already inverted—cool below, warm above. There is nothing new about the phenomenon. When Spanish explorer Juan Rodriguez Cabrillo first came upon the Los Angeles basin in 1542, he noticed that the smoke from Indian camps rose only a few hundred feet and then spread out horizontally, forming a kind of canopy. He named the nearby inlet the Bay of Smokes: It is now San Pedro Bay.

In other areas along the California coast, this low-lying layer of cool air—known to meteorologists as the marine layer—is dispelled by the air rising from the hot interior. But the San Gabriel and San Bernardino Mountains encircling the Los Angeles area protect the inversions there and ensure their longevity. Until well into the 20th Century, the inversion produced no ill effects. The area was big enough to absorb what pollution was generated, and often a horizontal land breeze pushed the contaminants out to sea in the evening. Smog first became noticeable during World War II, and residents immediately blamed the city's industries and refineries, which were operating at full throttle for the war effort. In the late 1940s, a campaign was launched to control industrial emissions. Pollution from these sources abated. But the smog remained.

Not only did the visible haze persist, but a variety of other puzzling effects also began to gain public notoriety. Residents complained about eye irritation on smoggy days, and local farmers began to point out extensive crop damage, which they believed to be smog related. In the early 1950s, Los Angeles County authorities called in Arie Jan Haagen-Smit, a biochemist and plant physiologist at the California Institute of Technology, to find out what was happening to the plants.

Haagen-Smit immediately realized that the crop damage—which consisted of an oily sheen on the underside of leaves—did not resemble the effects of known industrial toxins, such as sulfur dioxide. At first he tried exposing test plants to large doses of natural organic materials known to be present in the air. When this failed to reproduce the damage, he looked again at man-made pollutants, this time focusing on the abundance of nitrogen oxides and petroleum-based hydrocarbons revealed by chemical analyses of air samples.

Initially, fumigation of plants with nitrogen oxides and selected hydrocarbons brought negligible results, but Haagen-Smit soon realized that an important ingredient supremely characteristic of Los Angeles weather was missing—sunshine. To simulate the sun in the laboratory, he exposed the fumigated plants to ultraviolet light. After only a few hours, the test plants showed damage that was, he wrote, "indistinguishable from that noticed on plants exposed to smog."

Haagen-Smit quickly recognized the end product of the chemical reaction he had created: It was ozone, the highly reactive three-atom oxygen molecule. To confirm his findings, he conducted a simple test. It was well-known that ozone attacks the compounds in rubber, so the Caltech professor bent strips of rubber and subjected them to the Los Angeles smog. Under normal conditions, it takes about 45 minutes for stress cracks to appear where the rubber bends, but when exposed to the smog under ultraviolet light, the rubber cracked in six minutes. Only a remarkably high concentration—about 10 times the norm—could account for this result.

Los Angeles, he had discovered, was indeed experiencing a new kind of smog—technically not really smog at all because no visible smoke or fog was involved—that was every bit as dangerous as the traditional brew that had paralyzed London and Donora. And its source—revealed by the petroleum-based hydrocarbons used in the experiments—was that pervasive manifestation of American prosperity, the automobile. The chemical detective work that incriminated the beloved family car (and the petroleum refineries that nourished it) opened the way to an entirely new understanding of the workings of airborne pollutants.

Convincing others of the link between ozone and the cars that cruise the boulevards and freeways was a complicated task because, although nitric oxide is emitted from automobile tailpipes, ozone is not. But Haagen-Smit, soon to be known nationwide as "Dr. Smog," had shown that the linchpin

A Natural Trap for Tainted Air

The effects of air pollution are often worsened and extended by an atmospheric condition known as a temperature inversion. It occurs when a layer of warm air settles over low-lying cool air (the name of the phenomenon refers to the reversal of the normal progression to lower temperatures at higher altitudes). While it persists, air rising from the earth's surface loses its buoyancy at the overlying cap of warm air. Instead of maintaining its normal state of agitation, which usually disperses pollutants, the lower air stagnates, and pollutant concentrations increase.

Persistent inversions most often occur when warm air circulated aloft by a stable high-pressure system blankets a region. In some cities, geographical factors contribute to inversions so frequent and long-lasting that eye-stinging smog becomes a pervasive fact of life.

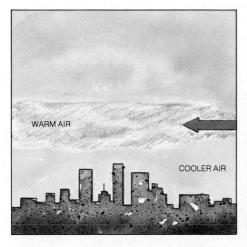

A layer of warm air circulated over Hartford, Connecticut, by a slow-moving high-pressure system *(red arrow)* traps cooler, polluted air beneath it. The exceptional buoyancy of a hot plume of steam carries it out of the trapped layer into the warm air above *(right)*.

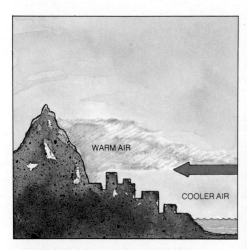

Cooled by the Pacific Ocean, air sweeping into the Los Angeles basin is often trapped by the surrounding mountains and warmer air aloft. This persistent, low-lying inversion fosters the city's well-known smogs.

Made dark and ominous by the setting sun, steam billows from an oil refinery in Edmonton, Alberta. In fact, many harmful pollutants escaping from refineries are invisible.

in the chemical transformation was sunlight, and doubters were convinced when scientists measured outdoor ozone levels around the clock: They found that the ozone level did indeed peak during the daytime and drop dramatically at night. The basic sequence of chemical reactions that creates smog was confirmed.

When an engine burns gasoline, many of the air's components escape combustion. Most of the oxygen is consumed, but the air's other principal ingredient, nitrogen, is left over. The extreme heat in the cylinders causes some of the nitrogen to combine with unburned oxygen to form nitric oxide, a colorless and mildly toxic gas molecule containing one nitrogen atom and one oxygen atom. If this gas remained unchanged in the air, it might be of little concern, but in the open it picks up another oxygen atom and becomes nitrogen dioxide, a toxic reddish brown gas.

Ultraviolet rays from the sun, absorbed immediately by the nitrogen dioxide molecules, split off an oxygen atom from each molecule and thereby re-create the nitric oxide gas. Meanwhile, a stray oxygen atom joins with a common two-atom oxygen molecule in the air to form the three-atom ozone molecule. This marriage is brokered by the hydrocarbons—unburned or partially burned gasoline vapors—which react chemically to promote ozone formation. As long as the car's engine is running, nitric oxide continues to be formed and the hydrocarbons in the exhaust facilitate the creation of ozone.

Ozone formation generally comes to a halt at the end of the day, but darkness does not necessarily bring respite from air pollution. Even without the transformation caused by ultraviolet rays, nitrogen dioxide in great quantities can foster all sorts of bronchial disorders, including emphysema. Ozone, however, is 15 times more potent. Tests have shown that even brief whiffs of it can induce headaches, drowsiness, nausea and the inability to concentrate, while sustained exposure can bring about injury to the respiratory system and impotency, among other effects.

In 1960 California passed the nation's first automobile-pollution-control laws. All new cars were required to be fitted with special emission-control systems that recycle unburned gases from the car's crankcase back into the cylinders. This recycling reduced the venting of carbon monoxide and unburned hydrocarbons. Later, other devices installed in the exhaust system further reduced harmful emissions. The catalytic converter, introduced in 1975, transforms exhaust gases into nitrogen, carbon dioxide and water—all harmless. The United States federal government followed California's lead in 1963 with the first of several Clean Air Acts, requiring similar controls nationwide.

After the new laws were put into effect, emissions from new automobiles were reduced, but two factors prevented any substantial improvement in air quality. For one thing, the number of cars on the road continually increased; in addition, many of the same potent gases and particles discharged by cars were released during some industrial processes. Total output of ozone-producing nitrogen dioxide seemed to hold steady for a few years, but after 1975, slight increases were again documented.

In addition to ozone-producing gases, automobiles emit many other toxins, including carbon monoxide. Some carbon monoxide is produced by car engines under all driving conditions, but in stop-and-go driving, when the engine burns inefficiently, the amount rises substantially. Even then, most of the gas disperses harmlessly, but if its dissipation is hampered by an

A schematic diagram of a pollution-control system shows how it reduces emissions from a coal-burning power plant. Flue gases are first passed through a precipitator *(upper left),* which collects electrically charged fly-ash particles on grounded plates. Then, in a scrubber *(lower left),* a spray of lime and water removes most sulfur dioxide fumes.

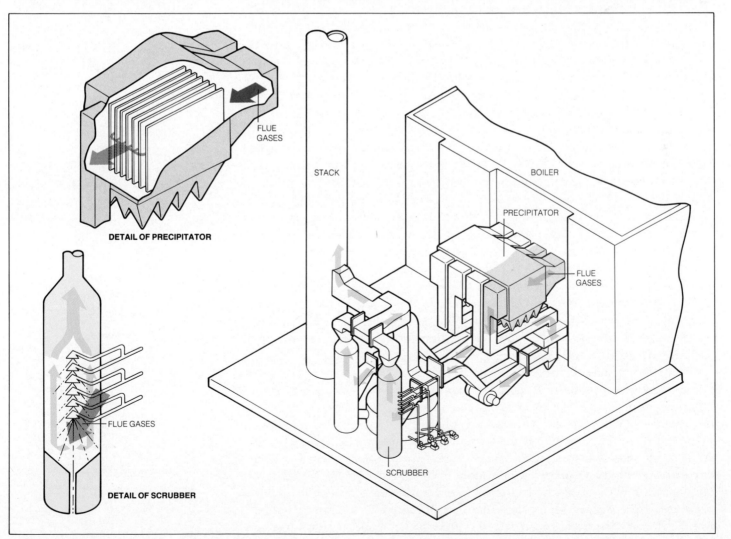

DETAIL OF PRECIPITATOR

FLUE GASES

DETAIL OF SCRUBBER

FLUE GASES

STACK

BOILER

PRECIPITATOR

FLUE GASES

SCRUBBER

inversion, or if its concentration is enhanced by heavy traffic, it can build up rapidly. Because carbon monoxide combines with some of the hemoglobin in the human bloodstream and prevents the blood from transporting oxygen, high concentrations can present a serious health hazard. Since 1972, emission-control systems have lowered the observed levels of carbon monoxide in American cities by one third.

Another noxious substance spewed from car exhaust is lead, which for many decades was used as a gasoline additive to reduce engine knocking and improve performance. When leaded gasoline is burned, millions of minute lead particles are sent into the air. These particles are so small that they can travel great distances on global winds; a significant number of them have even been found in the snow at the North and South Poles. An abundance of lead particles in the air can be harmful because, when inhaled, they enter the bloodstream and rupture red blood cells. Lead particles also damage cells in the gastrointestinal tract and the central nervous system, and cause brain damage in children. Recognizing the danger of unrestricted lead emissions, Congress passed a law that required the use of lead-free gasoline in all new cars, starting in 1975. Lead emissions from industry were also cut down by the installation of special scrubbing devices on smelter smokestacks. Once more, however, these potential gains have been offset by an overall increase in the number of cars and new industries: In the 1970s, total lead emissions increased from 130,000 tons per year to 200,000 tons.

Meanwhile, yet another kind of air pollution was brought to light. Ozone again was the focus, but this time the concern was depletion of the gas in the stratosphere rather than overproduction near the ground. Between six and 30 miles above the ground, stratospheric ozone functions to screen out harmful short-wave ultraviolet light. Even so, the ultraviolet light reaching the surface is strong enough to cause the production of protective pigments in human skin—a suntan. Excessive exposure, however, can cause damage to skin tissue; most often this takes the form of a painful sunburn, but in some cases it results in skin cancer. It has been estimated that a 1 per cent drop in the amount of stratospheric ozone could result in up to a 4 per cent increase in some forms of skin cancer. Concern about the ozone layer arises because ozone makes up only a tiny part of the upper atmosphere—about .005 of 1 per cent of the total volume of the stratosphere—and even minor depletions could have profound effects.

The excess ozone produced indirectly by the automobile at ground level cannot simply replenish the ozone layer in the stratosphere. Ozone is an extremely unstable substance, which, scientists believe, does not survive the trip to the stratosphere. Instead, it is constantly being created and destroyed independently at both levels.

The first widely noticed warnings about the potential hazards of ozone depletion were sounded in the 1970s, when several nations began producing supersonic commercial jetliners. Some scientists warned of dire consequences if the engines poured large amounts of nitrogen oxides directly into the stratosphere, where the craft were expected to cruise routinely. One such exhaust product, nitric oxide (one atom of nitrogen and one of oxygen), reacts readily with the triatomic ozone to produce nitrogen dioxide and a two-atom oxygen molecule. The nitrogen dioxide then reacts with a lone "free" oxygen atom to create two-atom oxygen and—once more—

nitric oxide. The cycle was thought to be virtually self-perpetuating. Early estimates predicted that a fleet of 500 supersonic airliners flown at an altitude of 12 miles would reduce the ozone in the stratosphere by some 16 per cent in five years. Nevertheless, France, Britain and the Soviet Union went ahead with their plans and began operating the planes on a regular basis.

Fortunately, later studies indicated that subsonic aircraft might produce enough ozone to make up for the losses. By the early 1980s, there was increasing evidence that subsonic airplanes flying near the tropopause create ozone in the same way that automobiles produce it at ground level, and some of this ozone rises into the stratosphere.

But even as the first objections were being voiced about aircraft ozone depletion, another threat to the ozone layer was perceived in, of all things, the aerosol spray can. Almost overnight, this seemingly insignificant device virtually attained the status of a doomsday machine because of the nature of the propellant used in most cans. Relied on for decades as a coolant in refrigerators and air conditioners, the chemical compound is called chloro-

Contrails blossom behind a fuel-tanker jet and four fighters cruising at 30,000 feet. The cloudlike ribbons, composed of water condensed from jet exhaust, also contain nitric oxide, which may affect the ozone layer.

fluorocarbon, although it is better known by its trade name, Freon. This compound is harmless to humans and is inert; that is, it does not react readily with other substances. But that very property was suddenly seen as a threat to the atmosphere.

When chlorofluorocarbon is released into the air, it evidently rises unchanged through the troposphere and arrives intact in the stratosphere. Laboratory experiments in 1973 suggested that a simple chemical sequence then takes place. The presumed steps run as follows: When exposed to intense ultraviolet light in the stratosphere, the compound breaks down to form chlorine oxide; this compound, in turn, acts as a catalyst in transforming ozone into common, two-atom oxygen; in the process of transforming the ozone into ordinary oxygen molecules, the chlorine is liberated to assault more ozone molecules; finally the chlorine combines with hydrogen to form hydrogen chloride, drifts down into the troposphere and is washed to the ground by precipitation.

Not all scientists accepted this scenario. Some said that the variables

For the most part, pollutants do not stay aloft in the atmosphere for long. In the troposphere, particles of soot and ash usually settle, or wash out in precipitation, within a few weeks; many gases are chemically broken down or are absorbed by moisture in two to four months. But when pollutants are injected into the stratosphere, they may remain suspended there for years.

Vast equatorial updrafts, part of the Hadley-cell circulation that drives the trade winds, are the main natural conduit into the stratosphere for trace gases and suspended particles. Among the most worrisome of the substances that follow this route through the tropopause (the upper boundary of the lower atmosphere) are chlorofluorocarbon—a propellant used in spray cans—and nitrous oxide emitted by chemically fertilized fields and the combustion of oil and coal. Both substances, say many scientists, may threaten the life-protecting ozone layer of the stratosphere.

But perhaps the heaviest polluters of the stratosphere are volcanic eruptions: Lofting an ash cloud laden with sulfur dioxide perhaps 12 miles, a major eruption can shroud an entire hemisphere in a veil of particles that reduces sunshine and lowers ground temperatures.

Once aloft, high-altitude pollutants are assured a long stay. Unruffled by the weather and vertical air mixing of the troposphere, the stratosphere is cleansed by only one circulation pattern. While strong east-west winds blow the air of the stratosphere around the globe, a languid horizontal drift gradually carries both ozone and pollution toward the Poles. High-altitude winds in the middle latitudes draw some air from the stratosphere downward into the troposphere, and the rest eventually sinks in the frigid polar areas, at last returning its freight of pollutants to earth.

A cross section of the atmosphere from Pole to Pole shows how air circulates in the troposphere and the stratosphere, dispersing ozone and pollutants. The figures in parentheses indicate the average length of time before fine particles lifted to various altitudes return to earth.

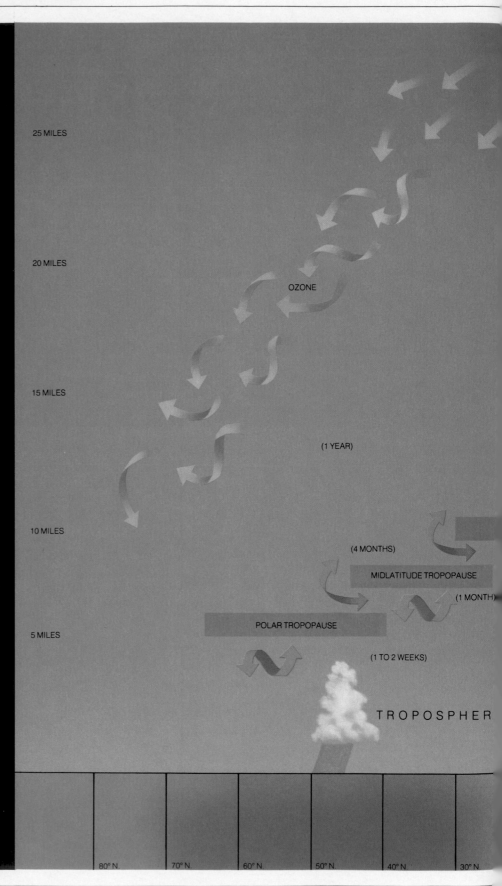

25 MILES

20 MILES

OZONE

15 MILES

(1 YEAR)

10 MILES

(4 MONTHS)

MIDLATITUDE TROPOPAUSE

(1 MONTH)

POLAR TROPOPAUSE

5 MILES

(1 TO 2 WEEKS)

TROPOSPHER

80° N.   70° N.   60° N.   50° N.   40° N.   30° N.

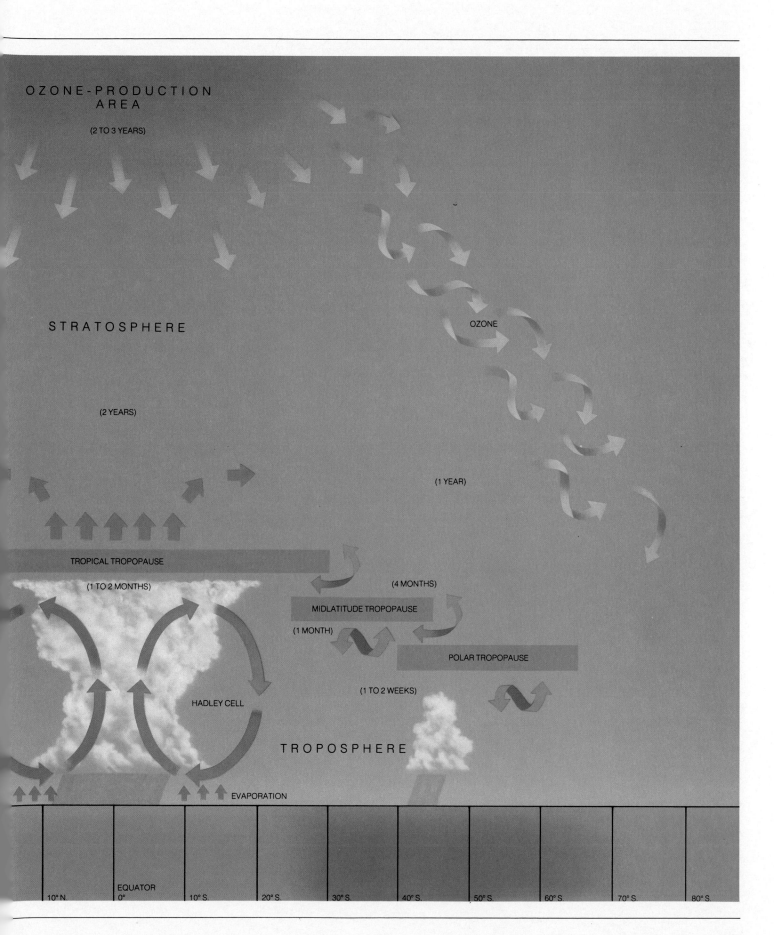

OZONE-PRODUCTION
AREA

(2 TO 3 YEARS)

STRATOSPHERE

OZONE

(2 YEARS)

(1 YEAR)

TROPICAL TROPOPAUSE

(1 TO 2 MONTHS)

(4 MONTHS)

MIDLATITUDE TROPOPAUSE

(1 MONTH)

POLAR TROPOPAUSE

(1 TO 2 WEEKS)

HADLEY CELL

TROPOSPHERE

EVAPORATION

| 10° N. | EQUATOR 0° | 10° S. | 20° S. | 30° S. | 40° S. | 50° S. | 60° S. | 70° S. | 80° S. |

involved are too numerous to warrant the assumption that laboratory results are applicable to the real atmosphere, and others felt that the speed of the ozone-depletion reactions envisioned is exaggerated. Pending further findings, the United States, alone among the many nations in which spray cans were in general use, imposed strict limits on the use of chlorofluorocarbon as an aerosol propellant; by the early 1980s, U.S. manufacturers had found that free nitrogen and carbon dioxide both worked well as substitutes.

While aerosol propellants were receiving such widespread attention, still another potential threat to the ozone layer was identified. In the 1970s and early 1980s, studies revealed that fertilizers used by farmers can also indirectly destroy high-altitude ozone. Nitrous oxide gas, released by bacterial action from nitrogen-enriched fertilizers, escapes into the air and rises into the stratosphere. There, ultraviolet light separates the nitrogen and oxygen in the compound, setting in motion chemical reactions that ultimately steal one atom of oxygen from the ozone, reducing it to normal, two-atom, molecular oxygen. Once again, scientists do not always agree on the precise mechanisms by which fertilizers release nitrous oxide or the extent of the damage caused in the stratosphere, but estimates made in 1982 at the National Center for Atmospheric Research in Boulder, Colorado, indicated that ozone depletion by nitrous oxide could be substantial by the 21st Century.

The most terrifying air-pollution specter arises in the form of the mushroom-shaped cloud that has become the symbol of the nuclear age. In addition to the incomprehensible immediate damage that would result from a wartime nuclear barrage, the airborne radioactivity would almost certainly prove lethal worldwide. Because nuclear fallout can remain in the stratosphere for several years, the world's winds would have ample time to distribute their deadly cargo to every continent. A chilling example of the way atmospheric circulation can distribute radioactivity occurred in 1964, when an American satellite carrying plutonium 238 as fuel accidentally burned up over the Indian Ocean; the resulting radioactivity was subsequently detected throughout both hemispheres.

Because radioactive particles can retain their potency for many years, predicting where and when harmful effects will be discovered is extraordinarily difficult. In one instance, when photographic film in Rochester, New York, was found to be fogged by radiation, the trouble was traced to the gelatin used in the manufacture of the film. The gelatin had come from Argentine cattle that had eaten grass contaminated by fallout from nuclear testing in the South Pacific and in Siberia.

Since 1963, when the United States, Britain and the Soviet Union agreed to halt aboveground testing of nuclear weapons, radioactive atmospheric pollution has declined dramatically, but many other windborne contaminants visit destruction on distant and unsuspecting victims. One such pollutant that gained notoriety in the late 1960s was acid rain.

The chief culprits in the case of acid rain were two familiar fossil-fuel byproducts, sulfur dioxide and nitrogen oxide. Produced by automobiles, power plants, smelters and a variety of other heavy industries, both substances combine readily with water vapor to make dilute acids. Nitric- and sulfuric-acid droplets can build up inside clouds and eventually precipitate

The air over the tall smokestacks of a coal-burning Tennessee power plant appears perfectly clean in a photograph *(right)* taken after pollution-control equipment had eliminated the visible emissions. But the use of an ultraviolet filter revealed *(above)* that hundreds of tons of sulfur dioxide—a source of acid rain—continued to erupt from the stacks each day.

out. In 1978 a storm in western Pennsylvania produced rain that was as acidic as lemon juice. Such high acidity can devastate lake fish and may also damage forests.

One of the great ironies of the acid-rain phenomenon is that its widespread effects are due, in part, to legislated efforts to reduce air pollution. To clean up the fouled air of neighborhoods near factories, regulations often required manufacturers to build tall smokestacks that would efficiently disperse the pollutants. Although this did indeed alleviate some local problems, it spread the waste products to previously uncontaminated areas. For example, a 1,250-foot-tall smokestack at a nickel smelter in Ontario, Canada, distributes thousands of tons of sulfur dioxide over vast downwind areas in eastern Canada and the Northeastern United States.

In both Europe and North America, some of the worst damage of acid rain has been wrought in wilderness areas, where mountain slopes trigger precipitation by forcing clouds to rise and cool. Unfortunately, the very remoteness of these areas delayed the discovery of the devastation. But in 1968, a Swedish soil scientist named Svante Odén was so shocked by the effects of acidic water in the lakes, soils and forests of Scandinavia that he accused the industrialized regions of England and central Europe of conducting "chemical warfare." His language caught the attention of the scientific community, and subsequent studies largely confirmed Odén's theory that industrial emissions were to blame for a rapidly worsening situation.

By the early 1980s thousands of Scandinavian lakes, once teeming with trout, had become virtually lifeless. Not long after Odén issued his report, scientists in the United States found a similar pattern downwind from industrial cities in the Midwest. And in the Northeastern states and eastern Canada, entire fish populations were being wiped out by the acid rain. Tests showed rainfall in these areas to be 10 to 30 times more acidic than uncontaminated rain. In the Adirondack Mountains in northeastern New York State, records indicated that the acidity of rain had increased 40-fold during the past half century. As it happens, the lakes in the Eastern United States are far more readily affected by acid rain than are most of those in the West, because the soil in the East is thin and the underlying rock is granite, an impermeable rock that cannot absorb or reduce acidity. In the Far West,

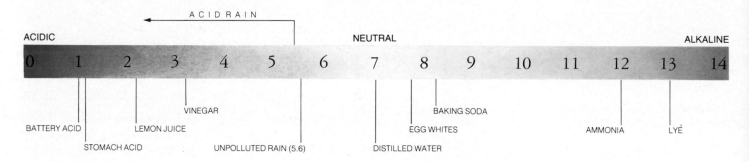

many lakes are protected by the presence of alkaline limestone layers and so suffer little from an increase in acidity.

In the vulnerable regions east of the industrial areas, much of the damage is done in the spring, when acids that have accumulated in the snow all winter long suddenly surge into lakes with the first meltwater. Fish suffer not only from the increased acidity, which inflames their gills, but from irritation by aluminum particles leached out of the soil by the acids. To ward off these toxins, fish typically produce a film of mucus on their gills, but often this defense mechanism is overtaxed: Mucus clogs the gills and the fish suffocate. Other equally hazardous metals leached from the soil and from lake-bottom sediments include mercury, lead and copper, which in high concentrations inhibit the hatching of salamander eggs, kill frogs and annihilate many forms of bacteria. Microscopic plants called phytoplankton may also die, signaling the obliteration of life by leaving the lake water crystal clear. "It's the ideal of a pristine lake," said one scientist. "I've been diving in some of those lakes and there's nothing left except a few water bugs."

The effect on the vegetation around a lake is less easily discerned. Laboratory experiments show that some plants are seriously harmed by acid rain, but there is also evidence that some trees actually flourish—for a time. In the early 1980s, a study made in Germany suggested that the reported increased tree growth is deceptive, because acid rain attacks trees in a three-stage assault. First, the nitrogen in the rain's nitric-acid component may act as a nutrient, enhancing growth for decades, before the acids begin to de-

A pH scale, which assigns the smallest numbers to the most acidic substances and the largest to the most alkaline, shows how acid rain compares with various other substances, including unpolluted rain. When the pH of water in lakes and streams drops much below 5 on the scale, most fish species die.

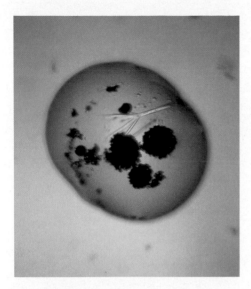

In this laboratory simulation of acid rain, needle-like sulfate crystals indicate the presence of sulfuric acid in a water droplet containing fly-ash particles and dissolved sulfur dioxide gas.

plete other vital nutrients in the soil. After some 40 years of acid rain, the soil becomes impoverished and tree growth slows. Then the sulfates from the acid rain become more devastating by combining with metals found naturally in the soils, especially aluminum. Aluminum sulfate, dissolved in the underground water, inhibits cell division in the roots, and eventually tree growth stops. The root damage also makes the trees vulnerable to a variety of diseases, and before long the starving tree succumbs to a combination of disease and poisoning. According to the study, acid rain killed trees covering some 3,750 acres in Bavaria between 1976 and 1981, and some 200,000 acres of forest land showed signs of extensive damage.

While no provable link between acid rain and human illnesses had been established by the early 1980s, some indirect associations seemed credible. For example, in Sweden it was found that increased acidity often causes well water to corrode copper plumbing pipes. In one case, the resulting increased copper content of well water is thought to have been responsible for some minor intestinal ailments. And a Swedish woman was shocked to find that her blond hair had turned green after she washed it in well water suffused with copper sulfate.

The problem of acid rain admits of no easy solution. Because the sources of pollution are often located far from the affected areas, it is extremely difficult to prove direct responsibility for the damage. Those industries that are suspected of contributing heavily to the acid-rain problem are reluctant to accept blame, and balk at the prospect of installing expensive pollution-control devices. The task of pinpointing sources is further complicated by the contribution of nitric oxide, which is produced by both industrial- and automobile-combustion processes. The travel of the pollutants across international borders is another complication, and ensuing legal tangles make it fairly easy to evade compulsory controls.

In Japan, considerable progress has been made. The Japanese government in 1968 enacted stringent controls on sulfur dioxide emissions and encouraged industries to switch to low-sulfur fuels. By 1975, sulfur emissions had been halved, and by the early 1980s, even stricter limits had placed Japan well ahead of other nations in sulfur-emission control.

If any good at all can be gleaned from the costs and complexities of urban pollution, the threats to the ozone layer, and acid rain, it might be the understanding that air pollution is no longer just a local, regional or even national concern. The coal and oil fires that power modern civilization cannot be extinguished without catastrophic social and economic effects. Yet evidence of the unhealthy and unlooked-for consequences of 20th Century progress continues to proliferate daily. And for all our scientific knowledge about the thin shell of air that protects the earth, the atmosphere may still conceal many surprises. Ω

AN AIRBORNE ASSAULT ON ANTIQUITY

"The past 40 years have done more corrosive damage to the art treasures of the Acropolis than all of the 2,400 years before." One observer's gloomy assessment of the situation in Athens applies wherever ancient sculpture and architecture have to coexist with industry and automobile traffic.

In Athens, as elsewhere, factories, oil-burning heaters and auto engines all spew sulfur dioxide and nitric oxide gases into the air until the laden miasma poses a threat not only to living things, but to stone itself. Rain cleanses the air, but only by turning the airborne gases into nitric and sulfuric acids. The resulting acid rain reacts with the marble of the Acropolis and transforms the stone into sugary-soft gypsum, which quickly washes away. In other European cities, monuments of limestone and sandstone, as well as marble, are wearing away under similar onslaughts of corrosive rain.

The acidic fumes can work their mischief without rain. Morning dew on stone can absorb corrosive gas from the air, as can moisture rising into a sculpture or a façade from damp ground. When wind and rain do not scour the corrosion away, gypsum and soot may combine into a hard, black crust while the stone beneath continues to erode.

By the mid-1970s, cultural authorities across Europe were alarmed at the extent of the damage. Statuary on the cathedral in Milan was crumbling; at the Acropolis, figures on Parthenon reliefs that had been visible just 10 years before were all but obliterated.

Along with local pollution-control efforts, some radical measures have been taken to salvage the threatened art: Some of the Acropolis sculptures now repose indoors, with cement copies taking their places in the open air. In Italy, restorers use sophisticated techniques to clean away corrosion, strengthen damaged stone and seal it against further damage. Even so, the restored monuments will need periodic cleaning as long as the polluted air of modern industrial society continues its assault on the artistic treasures of antiquity.

Surmounted by the 2,400-year-old Parthenon, Athens' besieged Acropolis looms through the airborne effluents of industry, homes and vehicles that threaten its ancient sculpture.

A cherub on the 19th Century façade of the
Milan Cathedral *(top)*, though partially
sheltered from the weather, is encrusted with
sooty corrosion mixed with pigeon droppings.
Washed by rain, a nearby minotaur *(above)*
is clean but badly eroded by the acid moisture.

A horse rears in a sculptured frieze on the west
face of the Parthenon, the pits and stains
that mar its surface bearing mute testimony to
the corrosive power of acid rain.

A healing poultice of clay covers a 16th Century Venetian statuette. The clay removes deposits from the pollution-damaged marble surface and helps to dry the internal moisture that had rendered the figure especially vulnerable to corrosive gases.

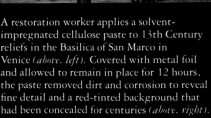

A restoration worker applies a solvent-impregnated cellulose paste to 13th Century reliefs in the Basilica of San Marco in Venice (*above, left*). Covered with metal foil and allowed to remain in place for 12 hours, the paste removed dirt and corrosion to reveal fine detail and a red-tinted background that had been concealed for centuries (*above, right*).

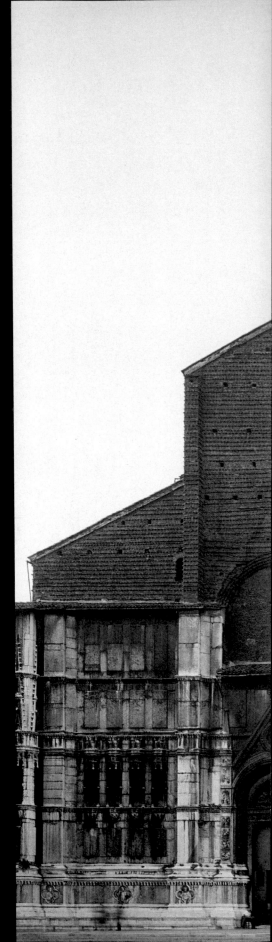

In a 1975 photograph taken during the restoration of the Church of San Petronio in Bologna, the right side of the façade, cleaned and sealed with a clear plastic resin, contrasts brightly with the untreated portals.

Chapter 6

PROPHECIES OF CLIMATES TO COME

During the 1970s a succession of unexpected and almost freakish weather events afflicted various parts of the globe. Bitter cold spread southward from the North Pole, and Arctic pack ice began drifting into temperate latitudes, posing a threat to shipping. Portions of the Soviet Union experienced their worst drought in centuries, while U.S. and Canadian grain-growing areas were beset by ruinously heavy spring rains. Year after year of extraordinarily scant rainfall in India and in the sub-Saharan region called the Sahel caused famines that claimed more than 200,000 lives. The winters of 1976-1977 and 1977-1978 brought murderous cold and snow to the Eastern United States. Snow fell in Miami, and in March 1978, Boston endured a "snow hurricane," an unusual blizzard with 100-mile-an-hour winds that dumped 27 inches of snow on the city in 24 hours. In 1979 all of the Great Lakes froze from shore to shore for the first time in memory. A European Common Market commission, announcing a major study of climate trends, pointed out that, during the previous 15 years, Europe had suffered the coldest winter since 1740, the driest winter since 1743 and the severest drought since 1726—but also the mildest winter since 1834 and the hottest month in 300 years.

With records tumbling everywhere, it seemed to many that something odd and perhaps drastic was happening to the world's weather. There were suggestions that a trend toward prolonged global cooling—perhaps even toward a new ice age—had begun; but others thought the trend was in the opposite direction, toward rising temperatures worldwide. Any such change would be of far more than academic significance. Even slight fluctuations in average temperature can affect the length of the growing season and, consequently, agricultural yields. Long-term variations in temperature also mean new patterns of energy use for heating and cooling. Since both food and fuel are potential sources of conflict among and within nations, making proper allowances for the role of climate in human events can be of critical importance.

Amid all the discussion of various alarming possibilities, climatologists—scientists who study long-term changes in weather patterns—remained notably circumspect in their interpretation of the wild fluctuations in the weather of the 1970s. For one thing, they had amassed enough data about past climate to know that erratic swings from drought to flood and from blizzard to heat wave have been common throughout the earth's history. In fact, climatologists are generally in agreement that, after having enjoyed an unusually favorable period of few meteorological extremes for

Researchers at the Goddard Institute for Space Studies in New York City scrutinize a computer display of world vegetation patterns for clues to their effects on climate.

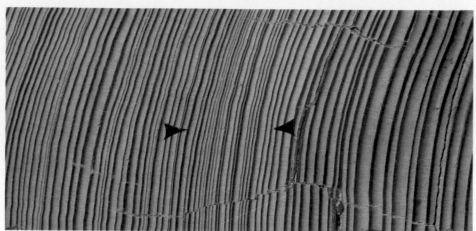

approximately 80 years, the earth is reverting to a more normal routine—such as that of the 1970s—in which natural excesses are likely to occur with unpleasant frequency.

During the era of benign weather that began around 1890, the world's climate was both warmer and more stable than anything mankind had experienced for almost 1,000 years, giving rise to the widespread impression that climate does not change much. This impression was partly the result of a lack of information about the long-term trends of the past. It had not been possible to make accurate measurements of weather conditions until the 17th Century, when the necessary instruments were invented, and for decades thereafter record keeping was spotty.

Nevertheless, modern climatologists have found ways to deduce the temperature and rainfall patterns of the distant past by studying historical records of such things as wine harvests, the annual blooming of cherry trees in Japan, and the growth and contraction of desert areas and alpine glaciers. In addition, they have measured the width of the annual growth rings of 5,000-year-old bristlecone pine trees, analyzed the chemical content of ice cores taken from the depths of the Greenland and Antarctic icecaps, pondered traces of pollen found in sedimentary rocks, examined the ancient remains of microscopic organisms in sea-floor sediments, and explored layers of silt and clay built up over the centuries on lake bottoms.

From such diverse data scientists have constructed a picture of repeated and extreme climatic fluctuation; in the past 700,000 years, there have been seven ice ages, interspersed with milder interglacial periods. The earth is now enjoying an interglacial that began some 11,000 years ago. During this period there have been a number of swings between moderate cold and warmth, some of them affecting different parts of the globe at different times. One warm spell, called the Big Climatic Optimum, peaked around 4000 B.C. and was followed by cooler millennia. The next warm period began about 500 B.C. and lasted more than 1,500 years (although a minor cold spell intervened from 700 to 800 A.D.). As the warming trend continued, Europe experienced the Little Climatic Optimum from about 800 to 1250 A.D.

This was the age of the Vikings, when Norsemen not only invaded northern Europe but expanded their territorial domain to encompass Iceland and Greenland; Leif Ericson and others are thought to have pressed even farther west to America. Greenland was named when its shores were indeed ver-

With the aid of a stereomicroscope, William J. Robinson of the University of Arizona reads an ancient climatic record chronicled in tree rings. The thin rings (*between arrows*) in a section of a Douglas fir were caused by a drought that lasted from 1276 to 1299 A.D.

A core sample taken from 300 feet beneath the Pacific Ocean floor near Mexico provides valuable information about the climate nearly 100,000 years ago. Each pair of light and dark bands represents one year of sedimentation: The light layers are the remains of microorganisms that flourished near the ocean surface during the dry season; the dark bands indicate silt that washed into the ocean from the land during the rainy season.

dant, and the Norse settlers were able to raise oats, barley and rye. The colony grew until it comprised some 3,000 settlers living on 280 farms. The Little Climatic Optimum was so warm that vineyards flourished in England, producing wines that supposedly rivaled those of France.

Around 1250, the balmy days of the Little Climatic Optimum began slowly but surely to wane. England once again became inhospitable to wine grapes, and on the Continent, vineyards that had flourished on hilltops had to be moved to lower, more protected sites. Long stretches of wet weather introduced a protracted cool period, and the particularly sodden decade of 1310 to 1320 brought terrible suffering to England and northern Europe. In some years wheat harvests were so poor that farmers did not even have enough grain to use as seed the following year. As a consequence, the price of wheat tripled, and death from starvation and related diseases reduced the population of some parts of England by two thirds.

Meanwhile, ice floes began to clutter the waters around Iceland, hindering access to Greenland. Soon, ships were unable to reach the Greenland colonists, farming became impossible and the settlement withered. In 1492 the Pope expressed concern that no bishop had been able to reach his Greenland flock for 80 years; he did not know that the last of the settlers had died by 1450. When a Danish archeologist began excavating a Norse cemetery on the Greenland coast in 1921, he found macabre evidence of the intensifying cold. The oldest remains were buried several feet below the surface, but later graves were shallower; evidently, the permafrost zone had moved upward, making deep digging impossible with primitive tools. The most recent graves—about 500 years old—were very close to the surface, which by 1921 had itself been hardened by permafrost.

The cold that turned Greenland into a frigid wasteland afflicted parts of Europe in the 15th Century. Temperatures began a decline that was not dramatic—they were, on the average, only 1° to 2° F. below those of 1200—but winters became longer and more severe, and summers were cooler and shorter. This climatic phase would last more than 300 years and would come to be known as the Little Ice Age.

With crops often damaged or destroyed by early-autumn frosts, food supplies were in some years inadequate, and famine again took a terrible toll. A vision of the era is seen in the work of the Flemish painter Pieter Bruegel the Elder, who portrayed cloudy, snow-clad landscapes filled with ice skaters plying the frozen canals and hunters making their way through the drifts. Ice floes drifted farther and farther south; sometime in the 14th Century a polar bear stepped ashore from a floe that had reached the Faeroe Islands, just 250 miles northwest of Scotland. The Thames River froze six times as often during the Little Ice Age as it had during the Little Climatic Optimum. Henry VIII took advantage of one of these occasions in 1536 to travel on the ice by sleigh from London downriver to Greenwich. In the Alps the glaciers crept down the mountainsides, often crushing houses that had been built in warmer times. As late as the 18th Century, the Little Ice Age was still in full cry. At one point during the American Revolution, Britain's fleet was immobilized by ice in New York Harbor; cannon could be slid across the harbor from Staten Island to Manhattan.

Asia was not spared the Little Ice Age. The cold settled over Japan and China in the 10th Century and lasted until the 14th Century, then moved westward to European Russia in the mid-14th Century and onward to central and western Europe in the 15th Century. Exactly why the large-scale climate shift traveled slowly around the globe in this manner rather than descending over the entire Northern Hemisphere at once remains a puzzle to climatologists.

Records from the Southern Hemisphere are sketchy, but it appears that, at times, the Antarctic pack ice actually receded as the Arctic's was advancing. This has led to the suggestion that climate trends in the two hemispheres may be out of phase with each other, one warming while the other cools. The evidence is far from conclusive and, in fact, other research indicates that the hemispheres experience similar climate changes, with Southern Hemisphere trends leading the changes in the north by one or two thousand years.

Very gradually, imperceptibly at first, the cold began to slacken; during the 19th Century, average temperatures increased a degree or two in the northern temperate latitudes. In the French Alpine village of Argentière, a glacier that had been pushing into the community's streets stopped in the 1850s, and soon began to retreat. Growing seasons lengthened and the Arctic pack ice retreated. The trend toward more benign weather lasted until World War II. Perhaps by coincidence but perhaps not, this was the heyday of Western imperialism, in much the same way that the Little Climatic Optimum had seen the Norsemen flourish; worldwide population surged upward, its growth sustained by bountiful harvests.

Around the middle of the 20th Century, temperatures in the Northern Hemisphere once more began to inch downward. Between 1940 and 1965 the Northern Hemisphere cooled by about .5° F., on the average, and from

Londoners frolic on the icy surface of the Thames River in this 1677 painting by Abraham Hondius. The Thames froze an unprecedented 10 times during the 17th Century, at the height of the period called the Little Ice Age.

The dramatic effect of a slow but steady warming trend can be seen in the retreat of the Argentière Glacier in the Alps. An engraving *(above, left)* shows the glacier near a French village in 1855; a photograph taken some 110 years later *(above, right)* reveals only distant traces of the glacier.

1951 to 1972 water temperature in the North Atlantic declined steadily. In parts of the United States, the trend was even more pronounced; average temperatures dropped by 1° to 2° F. in summer and 4° to 5° F. in winter. In 1974, climatologists George and Helena Kukla concluded a study of weather maps and satellite photographs of Arctic conditions during the previous seven years. They had discovered that the snow and pack-ice cover in the Northern Hemisphere had formed earlier, and had covered a larger area, in each of the last three years of the period than in the first four.

The Kuklas suggested that the meteorological extremes experienced in Europe and North America during the early 1970s might be linked to the increased snow cover. When the polar region grows colder in relation to the Equator, its cap of cold air tends to bulge farther southward than normal, developing more pronounced waves along its outer edge. These projecting lobes of cold air deflect the surface westerlies of the middle latitudes farther south, with consequences for climatic change that have been traced by a number of climatologists. Jerome Namias of the Scripps Institution of Oceanography proposed that the cold winters the United States had experienced during the 1970s could be attributed to a persistent, southward-projecting flow of cold polar air. Another scientist, Reid A. Bryson of the University of Wisconsin, argued that a similar kind of pattern was responsible for the devastating drought that struck Africa's Sahel region during the late 1960s.

Even in good years, rainfall in the Sahel is barely adequate to support farming and livestock raising. The African monsoon, which normally sweeps in from the South Atlantic across the nations on the coast, brings desperately needed moisture to the thirsty Sahel sands, much as the Indian monsoon normally drenches the lands south of the Himalayas. But in the late 1960s and early 1970s the African monsoon rains failed to reach the Sahel several years in a row; during the same period, the Indian monsoon weakened, and crops suffered in China. The failure of the African monsoon occurred, according to Bryson, because the deflection of the westerlies southward over Africa had displaced a large anticyclone that is an important determinant of local climate. This dry, high-pressure system normally hovers above the Sahara. When the monsoon encounters the southern edge of the anticyclone, its moisture condenses and falls as rain over the Sahel. But when the anticyclone is displaced southward, as was the case in the late

1960s and the early 1970s, the monsoon drops its burden of moisture before it reaches the Sahel.

Climatologists generally agree with the link Bryson proposed between the course of the westerlies and the drought in the Sahel. According to Bryson, the same mechanism—polar cold prompting a prolonged southward deflection of the westerlies—can be invoked to explain climatic variations in the past, such as the changes that began to affect much of the Northern Hemisphere around 1250, when Greenland grew colder and the Norse settlement went into decline. The drought that gripped the North American plains east of the Rocky Mountains at the same time may also have been the result of the southward-deflected westerlies; remaining over the region, they may have prevented moisture-laden tropical air from moving northward.

But the westerlies, and even falling polar temperatures, are only intermediaries and not initiators of climatic change. What could cause the Pole to cool in the first place? Bryson suggested that the drought in the Sahel—and, indeed, the global cooling trend that has only recently reversed—may ultimately have been triggered by dust. He argued that small particles in the atmosphere, especially pollutants and volcanic ash, can block a good deal of incoming solar energy while allowing terrestrial radiation to pass into space, thus reducing the amount of heat available to warm the earth. Most climatologists reserve judgment. The evidence, they say, is too meager to prove that dust was the cause of the drought in the Sahel.

In their investigations of climatic change, scientists divide the possible causes into those that are generated on earth—such as volcanic activity—and those derived from forces beyond the atmosphere, as from the sun. Of the possible earthly causes, a particularly worrisome one to some climatologists is the the so-called greenhouse effect, which traps the heat of solar radiation and warms the earth. The greenhouse effect is the result of the way two atmospheric gases, carbon dioxide and water vapor, act on solar and terrestrial radiation. Both of the gases are largely transparent to short-wave radiation, which includes ultraviolet and visible light, but they absorb long-wave, or infrared, radiation.

Most of the radiation reaching the earth from the sun is visible light, which passes through atmospheric carbon dioxide and water vapor and heats the earth's surface. (The very-short-wave solar radiation is largely absorbed by oxygen or ozone higher in the atmosphere.) Thus warmed, the earth reradiates much of the energy received—but in the longer-wave, infrared range. Some of this infrared radiation escapes into space, but some of it is absorbed aloft by carbon dioxide and water vapor, and heats the atmosphere. The warmed atmosphere, in turn, emits more infrared radiation—some back to earth and some out into space. The delicate balance between radiation gained and radiation lost by the atmosphere as a whole depends chiefly on the concentrations of the two gases. The total amount of water vapor present has apparently remained quite constant for as long as scientists have been able to measure it; the level of carbon dioxide, however, is a different matter.

Carbon dioxide is released into the air by a number of natural processes, among them the burning or decaying of plants and the exhaling of breath by animals and humans. (Animals absorb oxygen from air and exhale carbon dioxide, while plants take in carbon dioxide and expel oxygen.) Analysis of such things as the air trapped in ancient ice indicates that the concentration of carbon dioxide in the atmosphere remained fairly constant for thousands of years.

Since the 19th Century, however, the level has been rising. One contributing factor was the clearing of forests to meet the insistent demand for firewood and tillable land; more carbon dioxide was released by the burning and decaying of the wood, and fewer living trees remained to absorb the gas. Another factor was the increasing use of fossil fuels. Now, with industrial output and the use of automobiles on the rise, the amount of carbon dioxide in the atmosphere is going up at an increasing rate: Scientists estimate that it increased by about 20 per cent between 1880 and 1980, and they think it will rise by at least another 10 per cent by the year 2000, and by still another 10 per cent in the 10 years following.

Accumulation of carbon dioxide presumably causes more absorption of infrared radiation—and thus more heat—in the atmosphere. But estimates vary on just how much average global temperatures might be affected. Some scientists think that readings might be up by almost 2° F. by the first part of the 21st Century, with an even steeper rise thereafter—assuming that some other factor does not counterbalance the heightened greenhouse effect. Other computations, however, indicate that the temperature increase may not be that great.

The oceans could help. They absorb even more of the gas than trees and other green plants (which may be growing more luxuriantly these days because carbon dioxide is more plentiful). In the past, ocean waters have apparently been able to absorb 45 per cent or more of the carbon dioxide added to the atmosphere. However, the process is hardly straightforward: Any increase in temperature would not only diminish the oceans' capacity to absorb carbon dioxide, but would increase the rate of evaporation, and thus the amount of water vapor in the atmosphere. The greenhouse effect therefore could actually be heightened—although perhaps not right away: It would take a decade or two for a temperature increase to make itself felt in the upper layers of the ocean. Some of the keenest debates among climatologists arise when they try to assess the moderating role of the oceans.

Despite widespread concern about the greenhouse effect, scientists have no irrefutable proof that the increase in atmospheric carbon dioxide is in fact warming the earth. All predictions are based on laboratory studies and computer calculations. But a lack of conclusive data, as one climatologist has pointed out, "does not mean the theory is wrong."

The formidable difficulties of calculating long-term climatic trends are illustrated by a consideration of the effect of clouds. Depending on their type, they may either raise or lower temperatures at the earth's surface. Thin, wispy cirrus clouds, for instance, block relatively little incoming solar radiation but do inhibit the reradiation of heat from the earth, and thus tend to increase temperatures below. On the other hand, a low-lying canopy of stratus clouds can have the opposite effect: Since stratus clouds reflect more solar radiation and are also warm enough to radiate more heat back into space, they keep the earth's surface cool in the daytime.

Climatologists are similarly uncertain about the overall importance to climate trends of the particles that pollute the atmosphere. Large particles, such as dust or smoke from fires, have little effect on long-term climate

A Citadel of Atmospheric Studies

From a mile-high site in Boulder, Colorado, the National Center for Atmospheric Research (NCAR) conducts a full-scale scientific assault on the mysteries of the earth's atmosphere. Founded in 1960 by a group of American universities, and sponsored by the National Science Foundation, NCAR uses satellites, high-altitude balloons, research aircraft and a worldwide network of field stations to try to keep track of the atmosphere's every move.

Much of the data from these various sources sooner or later is digested by one great organizer—a huge, high-speed computer capable of performing 80 million operations per second. The computer's most notable feature is its use of mathematical models of atmospheric phenomena. For example, after the scientists have taught the computer the basic workings of a thunderstorm, they can feed it data on new conditions—such as increased wind speed or more moisture—and it will display the projected effects of the changes on the storm.

NCAR scientists have instructed the super-computer to peer both backward and forward in time. In one experiment, the computer constructed a model of the global climate during the age of the dinosaurs by mathematically relocating the earth's continents according to current theories of plate tectonics; the rearrangement of continents created additional open ocean—which would have retained more solar heat and helped maintain warmer temperatures around the world. In another experiment, the computer labored 60 hours to piece together a crude, 10-year projection of climatic trends; the results tend to confirm fears that increasing use of fossil fuels will cause a worldwide warming trend.

Built on a mountainside in Boulder, Colorado, the headquarters of the National Center for Atmospheric Research (NCAR) reflects the space-age technology applied within.

Environmental chemist Patrick Zimmerman checks a colony of termites. His studies have demonstrated that the digestive processes of the world's termites return more carbon dioxide to the atmosphere than is produced by modern consumption of fossil fuels.

A global weather map generated by the NCAR computer simulates the atmospheric dynamics of a single day. Isobars—lines of equal air pressure—delineate high- and low-pressure regions; the luminous areas represent clouds.

A specially designed telescope at NCAR's high-altitude observatory measures the sun's diameter each noon in an attempt to detect any shrinkage or expansion. The instrument is capable of spotting daily variations as small as 84 miles in the sun's 865,400-mile diameter.

A computer-simulated picture of a solar eclipse depicts the sun's corona (*green*) and magnetic-field lines looping out and radiating away from the sun's surface. Models such as this clarify the relationship between magnetic fields and the atmosphere.

because most of them rise only as far as the lower troposphere and settle back to earth quickly. Small particles, known as aerosols, are another story. Since they are so small, they remain aloft longer and are carried farther from their source by winds, thereby affecting larger areas. While large particles settle within hours or, at most, days, aerosols can stay in the upper troposphere for weeks and in the stratosphere for months or years.

Tropospheric aerosols can either cool or warm the earth's surface, depending on such factors as their size, chemical composition, shape and absorptivity. Stratospheric aerosols, however, tend to have a cooling effect on surface temperatures because they are too high up for the heat they gain by absorption to affect ground temperatures. No one yet knows when, or whether, the effect of man-made atmospheric aerosols might become truly significant in terms of climatic change.

The one type of atmospheric aerosol that scientists generally agree does produce some cooling is that spewed out of volcanoes. The dense clouds of dust, ash and gas rising from volcanic eruptions can reach into the stratosphere, travel around the world and linger for years, reflecting into space solar heat that would normally warm the earth. The British climatologist Hubert Lamb has gathered evidence that outbreaks of volcanic activity tend to coincide with cold periods. He found, for example, that there was an unusually high frequency of eruptions between 1500 and 1900, roughly the same time span as the Little Ice Age.

But the long-term effects of massive volcanic eruptions are, again, open to dispute. After years of intensive study, the Soviet climatologist M. I. Budyko concluded that a severe eruption yielding a particularly dense cloud could reduce the amount of direct solar radiation reaching the earth's surface by 10 per cent for one to two years. Allowing for the different ways in which land and water absorb or reflect heat and for the circulation and dissipation of heat by the atmosphere, Budyko calculated that such an eruption could reduce temperatures the world over by as much as 1° F. The 1963 eruption of Mount Agung in Bali, for instance, is believed to have been responsible for an average decline of almost 1° F. in the tropics.

A few climatologists think that a long series of eruptions could conceivably initiate a full-scale ice age, but that scenario does not enjoy much scientific support. "There is no evidence," states meteorologist Stanley Gedzelman, "that volcanic eruptions have ever produced any long-lasting climate changes. The only way for volcanic eruptions to do so would be if the level of volcanic activity was at least 10 times greater than it has been over the last century."

To identify forces that could produce long-term changes, many climatologists look outside the intricate, generally self-stabilizing and self-adjusting atmosphere, and consider what might be wrought by the sun or by the earth's relationship to the sun. For if the entire atmosphere runs on energy coming from the sun, perhaps major climatic changes occur only if that energy supply fluctuates. The total amount of radiation reaching the earth from the sun is called the solar constant—a name that reflects the long-standing belief that it never changes. Evidence that began to emerge in the late 19th Century suggested that the assumption needed reexamining, and the clues were provided by sunspots.

These dark blotches on the surface of the sun, now known to be tempo-

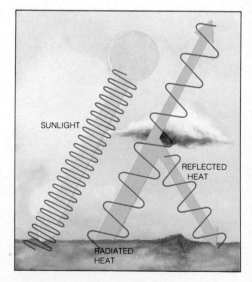

The greenhouse effect, shown here in simplified form, occurs when carbon dioxide and water vapor in the lower atmosphere reflect heat that would otherwise be radiated out into space by the sun-warmed earth.

The steady increase of carbon dioxide in the air is clearly documented in this chart of readings taken at an atmospheric-research station in Hawaii. The carbon dioxide level fluctuates widely each year because plants absorb the gas during the growing season, then release it when they decay in the autumn and winter.

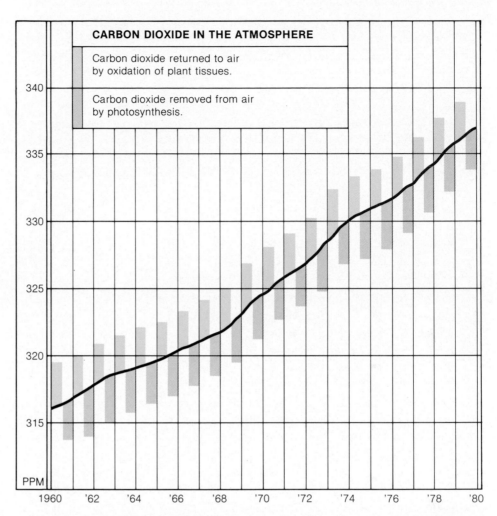

CARBON DIOXIDE IN THE ATMOSPHERE

Carbon dioxide returned to air by oxidation of plant tissues.

Carbon dioxide removed from air by photosynthesis.

rary intensifications of the solar magnetic field, were studied by Galileo in 1611 (he used them to calculate the speed of the sun's rotation) and had been noticed centuries before that. But by the mid-19th Century, just about the only thing that was known about them was that the outbreaks reach a peak approximately every 11 years. Then the British astronomer E. Walter Maunder, combing through old records of sunspot sightings, found that for a period of 70 years—from 1645 to 1715—almost no sunspots had been observed. They had shown up regularly before that, and they resumed thereafter. Maunder's writings about this odd gap drew little attention until 1976, when U.S. climatologist John Eddy pointed out that the interlude, which he named the Maunder Minimum, coincided with a particularly severe period of the Little Ice Age.

Other scientists soon found more evidence of a connection between sunspot cycles and changes in climate. They discovered that sunspots in fact have a double cycle: Their frequency reaches a maximum about every 11 years, but the polarity of the sun's magnetic field changes over a 22-year period. Climatologists studying tree rings, which are narrower in years of light rainfall, learned that droughts had recurred in the Western plains region of the United States approximately every 22 years. Other researchers found that, over the course of nearly five decades, the water level of Lake Victoria in Africa had risen and fallen in synchronization with the 11-year cycle. By the late 1920s, however, any connection that may have existed

A solar prominence of flaming gases licks 370,000 miles into space, preceded by a coronal transient—a ballooning of the sun's outer corona—visible in this composite photo. When coronal transients reach the earth's magnetic field, they trigger enormous magnetic storms.

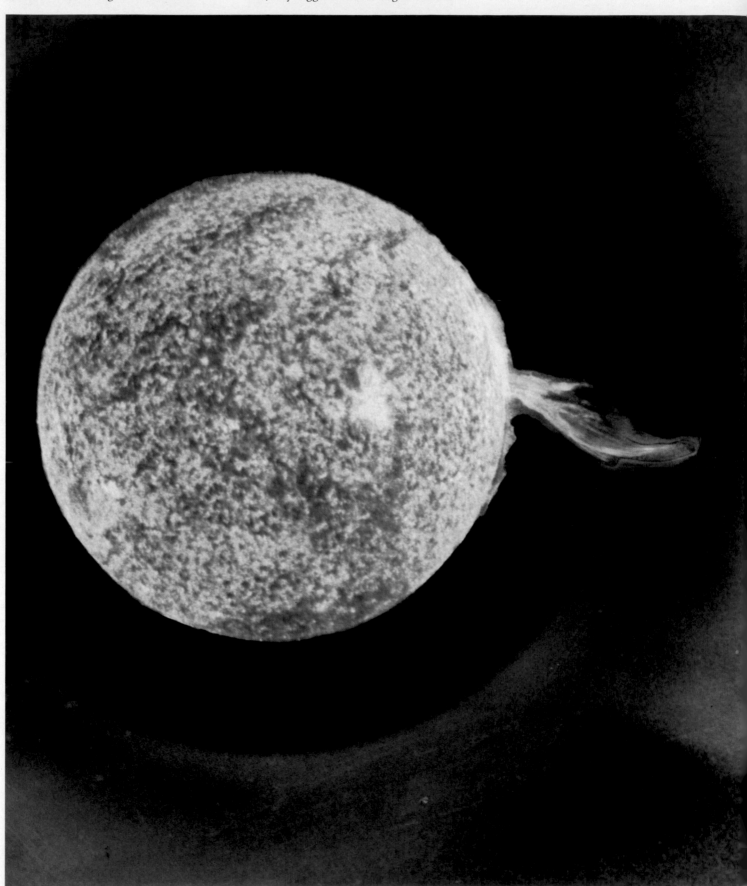

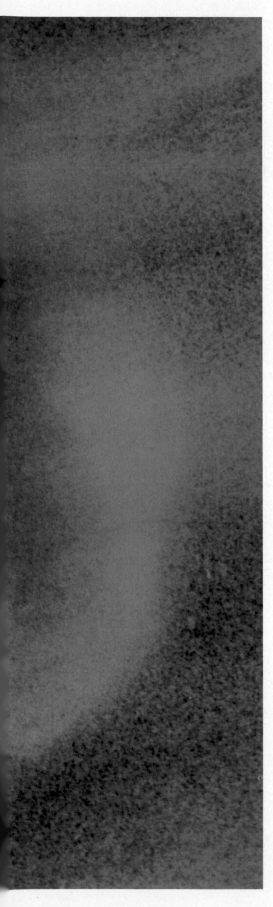

between sunspots and the lake's water level was broken: The lake went completely out of phase with the 11-year cycle.

Climatologists believe it is possible that sunspots influence climatic change—though how, they cannot say. Spacecraft ranging far from earth have confirmed that sunspot activity affects the solar wind, the stream of electrically charged particles emitted by the sun. The earth's magnetic field deflects most of these particles, but some manage to enter the upper atmosphere, where they collide with gas molecules and emit the shimmering lights of the aurora. In the early 1950s the American meteorologist Walter Orr Roberts made what seemed at the time to be a farfetched connection. Whenever the aurora borealis illuminated the winter skies above Alaska with unusual brilliance, he discovered, the low-pressure storm systems in the region became especially vigorous. His observations were confirmed, but after 1973 the mysterious relationship disappeared. No one has been able to explain either its existence or its absence.

Another possible extraterrestrial determinant of global climate operates so slowly that it could not cause such minor shifts as a drought or even a Little Ice Age. However, scientists believe it may be the principal mechanism of ice ages and warm periods thousands of years in duration. According to a hypothesis advanced in the 1920s by the Yugoslavian geophysicist Milutin Milankovitch, changes in the earth's spatial relationship to the sun can bring about profound climatic changes simply by varying the amount and geographical distribution of solar radiation.

Milankovitch noted that as the earth spins and moves around the sun, both its orbit and its attitude change slightly. The orbit varies from almost circular to strongly elliptical and back again every 93,000 years or so. The earth's tilt in relation to the plane of its orbit—the cause of earthly seasons—changes from about 22 degrees to more than 24 degrees and back every 41,000 years. The earth also wobbles, rocking in a circular motion around its axis like a slowing top, and this too has a cycle: One full wobble consumes 25,800 years. By altering the distance between the sun and earth or changing the angle at which radiation strikes particular points on the earth, these moves alter the amount of solar energy reaching certain latitudes in certain seasons.

Evidence that supports the critical role Milankovitch attributed to these cycles has accumulated steadily. For instance, scientists from the Lamont-Doherty Geological Observatory in New York discovered that variations in the type of oxygen and in the distribution of the remains of minute marine life—found in sedimentary samples taken from the floor of the Indian Ocean—indicate periodic and severe climate changes. The sea-core record suggests that some of these changes have peaked every 23,000 years, others every 41,000 years and still others about every 100,000 years. It seems highly unlikely that the similarity to the span of orbital variations is mere coincidence.

Because three factors are involved—orbit, tilt and wobble—their cumulative warming or cooling effects are extremely difficult to calculate. Even considered singly, the climatic effect is not easy to predict. Tilt, for example, affects the difference between winter and summer more than it affects the average global temperatures. For the past 10,000 years or so, the degree of tilt has been lessening, a process that should theoretically produce warmer winters but cooler summers. Warmer winters would mean more snowfall at the Poles; cooler summers would mean that less of the accumulated snow would melt each year, and more would remain to form glacier ice. After a period of years the polar ice sheets would increase significantly and eventually usher in a new ice age. Despite uncertainties about some details of the scheme, many climatologists are convinced that the Milankovitch model largely explains the recurrence of ice ages throughout the last million years or so of the earth's history. And if the explanation is correct, it is almost inevitable that the earth will once again experience another ice age, perhaps within the next several thousand years.

Whether the earth is headed toward cooler or warmer times in the next several decades obviously cannot be decided with any degree of certainty. After the decline in global temperatures charted from the 1940s to the mid-1960s, the averages once again started inching up. During the same years George and Helena Kukla were reporting an increase in Arctic ice, the Antarctic ice pack was shrinking. Many climatologists are of the opinion that whatever trend is operating at the moment, the intensified greenhouse effect will probably overwhelm any cooling trend during the next half century or longer.

Two of the most respected climatologists in the United States—William W. Kellogg and Stephen H. Schneider of the National Center for Atmospheric Research in Boulder, Colorado—have concluded "that mankind is likely to be warming the earth, and that the global climate change as early as the turn of this century could well be larger than any of the natural

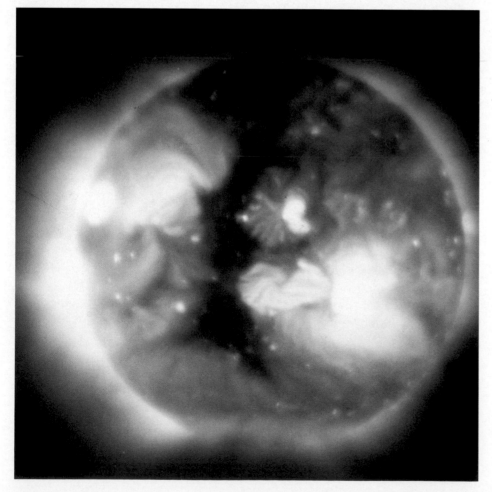

The dark blotches on this X-ray image of the sun are low-density holes in the sun's corona. From these holes erupts the invisible stream of charged particles known as the solar wind.

climate changes that we have experienced in the past thousand years or more. This will lengthen the average growing season in many places, and shift the rainfall patterns as well—some for better, some for worse." William Elliott of the National Oceanic and Atmospheric Administration put it less formally: "If I had to make a prediction for the year 2025, I would say that it will be warmer than today. But I could be wrong, of course."

Despite the uncertainty, advances in scientific knowledge about the atmosphere have, inevitably, spawned schemes to control the climate. Though such proposals ordinarily have some apparently laudable end, such as increased harvests, they all hold the possibility of unpredictable and irreversible damage to the earthly environment and to society as well. "Before we take comfort in our growing ability to bend nature to our purposes," Stephen Schneider has written, "we must remember that the atmosphere, the oceans, the land surfaces, and the snow and ice fields—which are the major components of the climate system—all act in concert to determine the climate. The forces that generate winds and rain at one particular place on earth are coupled in varying degrees to those forces at places on the other side of the earth, a relationship meteorologists call teleconnections. Every place on earth is connected to some extent by the climate system to every other place." Thus, as Reid Bryson has suggested, plunging temperatures at the North Pole can trigger monsoon failures in Africa and India and, simultaneously, drought in North America and endless rain in Europe.

A number of examples of what human interference with the earth's teleconnections might bring are afforded by some recent large-scale engineering projects. Distressed by the low productivity of much of the grain-growing region in Siberia and vexed by the relentless Siberian cold, the government of the Soviet Union designed a plan to dam the Ob and Yenisei Rivers, which flow northward through Siberia to the Arctic Ocean, to create several large reservoirs for the irrigation of Siberian farmland. These artificial inland seas would also, it was thought, warm and moisten the winds blowing across them to the dry, frigid steppes beyond. Warmer temperatures and increased rainfall might well make previously barren steppe land fertile. Another part of the plan involved impounding several other rivers in north Russia and channeling some of the water to replenish the Caspian Sea, which has been shrinking in size and growing steadily saltier, thus harming caviar production.

Aside from the huge cost of the project, some analyses suggested serious potential problems. The resulting change in temperature and humidity could alter the Siberian high-pressure system, which exercises a strong influence on the weather in China and much of southern Asia. A chain reaction could affect not only the annual monsoons in Asia but conditions even farther away, such as rainfall in California and New York. Furthermore, depriving the Arctic Ocean of the fresh water contributed by the two rivers could impede the formation of sea ice, since the increased salinity would lower the freezing temperature of Arctic waters. A reduction in the expanse of ice would raise the temperature and the humidity in the Arctic sufficiently, said some climatologists, to increase the annual snowfall there—with still further unpredictable repercussions. Many Soviet scientists disagree with these scenarios and maintain that if the project is ever completed, it will affect only the Soviet Union.

Another scheme proposed by the U.S.S.R.—and reportedly even discussed by President Gerald Ford and Soviet Chairman Leonid Brezhnev at a 1974 meeting in Vladivostok—called for the construction of a dam across the 60-mile-wide Bering Strait between Siberia and Alaska. At first the idea seemed attractive. If warm Pacific water were kept from mixing with—and being chilled by—the cold Arctic Ocean, Siberia's eastern shore and parts of Alaska would be warmer, ports currently unusable in winter would be ice-free and agriculture would spread farther north. But the reverberations would likely have been felt far away. If the dam allowed no water to pass into the Pacific, the icy polar waters would have to go in another direction and might drain across the top of North America and down Canada's east coast, further cooling the already forbidding Labrador Current and shortening the growing season in Canada's Maritime Provinces so much that agriculture might become nearly impossible.

The planners envisioned drawing the warm Gulf Stream northward into the Arctic basin to melt the polar pack ice. No one knows precisely what effects the melting of the ice would have on climate, but it is easy to imagine what might happen to the climate of England or Scandinavia, if the Gulf Stream was to desert its usual northeasterly course.

"At the present time, most meteorologists would agree that warming the Arctic by a small amount would change the global weather and climate," wrote meteorologist and author Louis J. Battan in 1969. "Unfortunately, nobody knows *how* it will change. Will the deserts be brought into bloom and the swamps dried up? Or will the swamps be even more inundated and deserts enlarged? Will the farming regions of the world get more or less rain and snow and when will it fall? Will the ocean surface rise and flood the low-lying cities of the world? Will the changes in general circulation initiate another ice age? No one yet knows the answers to these and a great many other related questions. Until we do, or at least can take a good guess, we better be careful tinkering with the global atmosphere." **Ω**

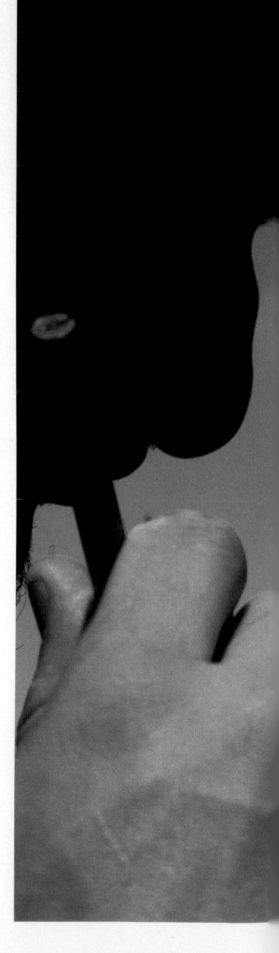

As if gazing into a scientific crystal ball, a researcher at an Australian meteorological station watches a parheliometer record information that might help to reveal climatic trends. Focused by the glass sphere, the sun's rays burn a record of each day's solar radiation into calibrated cards mounted around it.

ACKNOWLEDGMENTS

For their help in the preparation of this book the editors wish to thank: **In France:** Boulogne-Billancourt—Jacques Dettwiller, Press Attaché, Météorologie Nationale; Paris—Maxime Préaud, Curator, Bibliothèque Nationale; Comte Louis de Chabrol; Gérard Baschet, Éditions de l'Illustration; Pierre Berthon and Bruno Jammes, Institut de France; Marthe de Rohan-Chabot; Lucien Scheler; Saint-Maur-des-Fossés—Guy Penazzi, Institut National d'Astronomie et de Physique. **In Great Britain:** Bracknell, Berkshire—The National Meteorological Office; R.A.S. Ratcliffe, Royal Meteorological Society; Glasgow—Dr. D.A.R. Simmons; The Isle of Wight— R. K. Pilsbury; London—Francis Dunkels, British Museum; Library, Institute of Geological Sciences; Celina Fox, Museum of London; I. M. McCabe, Library, Royal Institution of Great Britain; Library, The Royal Society; Martin Andrewartha and Martin Evans, Science Museum; Kathryn Carver, Science Museum Library. **In Greece:** Athens—Committee for the Restoration of the Akropolis. **In Italy:** Bologna—Ottorino Nonfarmale; Raffaella Rossi Manaresi, Centro "Cesare Gnudi" per la Conservazione delle Sculture all'Aperto; Florence— Mara Miniati and Franca Principe, Istituto e Museo di Storia della Scienza; Milan—Carlo Ferrari da Passano, Veneranda Fabbrica del Duomo; Rome—Giovanni Urbani, Director, and Rita Cassano, Istituto Centrale del Restauro; Vatican City—Reverend George Coyne, Director, Specola Vaticana; Venice—Margherita Asso, Superintendent, and Mario Piana, Soprintendenza per i Beni Ambientali e Architettonici; Maria Teresa Rubin de Cervin, Director, and Tiziana Cipelletti, UNESCO. **In Japan:** Tokyo—Dr. Takasi Oguti, Geophysics Research Laboratory, University of Tokyo. **In the Netherlands:** Leyden—Prentenkabinet, Leyden University. **In the United States:** Alabama—(Huntsville) Dr. Valerie Neal, Essex Corporation; Alaska—(Fairbanks) Al Belon, Charles S. Deehr and Glenn Shaw, Geophysical Institute, University of Alaska; Arizona—(Tucson) Dr. Louis J. Battan, Professor of Atmospheric Sciences, University of Arizona; California—(El Monte) Dr. Eric Lemke; (Pasadena) California Institute of Technology;

(Studio City) Margaret Brunelle; Colorado—(Boulder) Joan V. Frisch, Dr. Douglas Hoyt, Dr. William W. Kellogg, Dr. Gordon Newkirk and Dr. Stephen H. Schneider, National Center for Atmospheric Research (NCAR); Dr. Ronald Holle, Office of Weather Research and Modification, National Oceanic and Atmospheric Administration (NOAA); (Fort Collins) Dr. Elmar Reiter, Atmospheric Sciences Department, Colorado State University; (Golden) Robert Noun, Solar Energy Research Institute; District of Columbia—Dr. John C. H. Wang, Federal Communications Commission (FCC); Dr. George Esenwein and Dr. Robert D. Hudson, Environmental Observation Division, Dr. John Lynch and Dr. Michael Wiskerchen, Office of Space Science and Applications, National Aeronautics and Space Administration (NASA); Dr. Francis Johnson, National Science Foundation; Bruce H. Needham, Chief, Satellite Data Services Branch, and Dr. Vince Oliver, NOAA; Robert Werner, Weather Service Forecast Office; Hawaii—(Hilo) Dr. Kinsell Coulson and Dr. Thomas de Foor, Mauna Loa Observatory, NOAA; Illinois— (Champaign) Dr. Wayne M. Wendland, Head, Climatology Section, Illinois Department of Energy and Natural Resources; Iowa—(Iowa City) Dr. James A. Van Allen, Department of Physics and Astronomy, University of Iowa; Maryland—(Annapolis) Greg Harlin and Rob Wood, Stansbury, Ronsaville, Wood Inc.; (Bowie) I'Ann Blanchette; (Gaithersburg) Carol Schwartz; (Greenbelt) Nelson W. Spencer, Chief, Laboratory for Planetary Atmospheres, and Dr. Jay Zwally, Ice Research Section, Goddard Space Flight Center, NASA; (Potomac) Fred Bigio; (Rockville) Dr. James Angell, Dr. William Elliott, Research Meteorologists, Dr. Lester Machta, Director of Air Resources Laboratories, and Dr. J. Murray Mitchell, Senior Research Climatologist, NOAA; (Silver Spring) Walt Cottrell, Public Affairs Office, NOAA; Massachusetts—(Cambridge) Randall Dole, Dr. Edward Lorenz and Dr. Reginald Newell, Department of Meteorology and Physical Oceanography, Massachusetts Institute of Technology; Michigan— (Kalamazoo) Dr. Philip Micklin, Department of Geogra-

phy, Western Michigan University; New Hampshire—(Gorham) Greg Gordon, Mount Washington Observatory; New Jersey—(Princeton) Dr. Abraham H. Oort, Geophysical Fluid Dynamics Lab, NOAA; New York—(Albany) Dr. Roger J. Cheng, Atmospheric Sciences Research Center; (Palisades) Dr. William Donn, Lamont-Doherty Geological Observatory; North Carolina—(Raleigh) Dr. Ellis Cowling, School of Forest Resources, North Carolina State University; (Research Triangle Park) Dr. Kenneth L. Demerjian, Division of Meteorology Environment Science Laboratory, Environmental Protection Agency (EPA); Ohio—(Cleveland) Daron Boyce, Great Lakes Ice Forecaster, National Weather Service; Oklahoma—(Bartlesville) Mr. and Mrs. Donald E. Carr; Pennsylvania—(Maytown) Ken Townsend; (University Park) Dr. Lynn Carpenter, Ionospheric Research Laboratory, Dr. Alistair B. Fraser, Professor of Meteorology, and Dr. John Olivero, Associate Professor of Meteorology, The Pennsylvania State University; Virginia—(Alexandria) Walter Hilmers Jr.; (Vienna) Kathy Rebeiz; Wisconsin—(Milwaukee) Dr. Robert Greendler, Professor of Physics, University of Wisconsin. **In West Germany:** Berlin—Dr. Roland Klemig and Heidi Klein, Bildarchiv Preussischer Kulturbesitz; Wolfgang Streubel, Ullstein Bilderdienst; Bonn—Dr. Helmut Kraus, Direktor, Inge Lockwood, Meteorologisches Institut der Universität; Heidelberg—Dr. Hermann Schildknecht, Direktor, Institut für Organische Chemie der Universität Heidelberg.

Particularly useful sources of information and quotations used in this volume were: *The Atmosphere* by Richard A. Anthes, John J. Cahir, Alistair B. Fraser and Hans A. Panofsky, Charles E. Merrill, 1981; *World of the Wind* by Slater Brown, Bobbs-Merrill, 1961; *The Edge of Space* by Richard A. Craig, Doubleday, 1968; *The Science and Wonders of the Atmosphere* by Stanley David Gedzelman, John Wiley & Sons, 1980; "The Global Circulation of Atmospheric Pollutants" by Reginald E. Newell, *Scientific American,* January 1971.

The index was prepared by Gisela S. Knight.

BIBLIOGRAPHY

Books

Anthes, Richard A., John J. Cahir, Alistair B. Fraser and Hans A. Panofsky, *The Atmosphere.* Charles E. Merrill, 1981.

Battan, Louis J.:
Cloud Physics and Cloud Seeding. Doubleday, 1962.
The Unclean Sky. Greenwood Press, 1980.

Beer, Tom, *The Aerospace Environment.* Springer-Verlag, 1976.

Bentley, W. A., and W. J. Humphreys, *Snow Crystals.* Dover Publications, 1962.

Bernarde, Melvin A., *Our Precarious Habitat.* W. W. Norton, 1970.

Blumenstock, David I., *The Ocean of the Air.* Rutgers University Press, 1959.

Boyer, Carl B., *The Rainbow: From Myth to Mathematics.* Thomas Yoseloff, 1959.

Breuer, Georg, *Air in Danger: Ecological Perspectives of the Atmosphere.* Transl. by Peter Fabian. Cambridge University Press, 1980.

Brown, Lloyd, *The Story of Maps.* Little, Brown, 1949.

Brown, Slater, *World of the Wind.* Bobbs-Merrill, 1961.

Bryson, Reid A., and Thomas J. Murray, *Climates of Hunger: Mankind and the World's Changing Weather.* University of Wisconsin Press, 1977.

Calder, Nigel, *The Weather Machine.* Viking Press, 1974.

Carr, Donald E.:

The Breath of Life. W. W. Norton, 1965.
The Sky Is Still Falling. W. W. Norton, 1982.

Cole, Franklyn W., *Introduction to Meteorology.* John Wiley and Sons, 1980.

Craig, Richard A.:
The Edge of Space. Doubleday, 1968.
Upper Atmosphere: Meteorology and Physics. Academic Press, 1965.

Critchfield, Howard J., *General Climatology.* Prentice-Hall, 1974.

Crowther, James G., *Founders of British Science.* London: Cresset Press, 1960.

Dobson, G.M.B., *Exploring the Atmosphere.* Oxford: Clarendon Press, 1963.

Eddy, John A., *A New Sun: The Solar Results from Skylab.* National Air and Space Administration, 1979.

Flohn, Hermann, *Climate and Weather.* Transl. by B. V. de G. Walden. McGraw-Hill, 1969.

Frisinger, H. Howard, *The History of Meteorology: To 1800.* Science History Publications, 1977.

Galilei, Galileo, *Two New Sciences.* Transl. by Stillman Drake. University of Wisconsin Press, 1974.

Gallant, Roy A., *Earth's Changing Climate.* Four Winds Press, 1979.

Gedzelman, Stanley David, *The Science and Wonders of the Atmosphere.* John Wiley & Sons, 1980.

Glaisher, James, *Travels in the Air.* Lippincott, 1871.

Gnudi, Cesare, et al., *Jacopo Della Quercia e la Facciata di*

San Petronio a Bologna. Bologna: Centro per la Conservazione delle Sculture all'Aperto, 1981.

Greenler, Robert, *Rainbows, Halos, and Glories.* Cambridge University Press, 1980.

Gribbin, John:
Forecasts, Famines and Freezes. Walker and Company, 1976.
Future Weather and the Greenhouse Effect. Delacorte Press/Eleanor Friede, 1982.
Weather Force: Climate and Its Impact on Our World. Putnam, 1979.

Heuer, Kenneth, *Rainbows, Halos and Other Wonders.* Dodd, Mead, 1976.

Howard, Luke, *Seven Lectures on Meteorology.* London: Harvey and Darton, 1843.

Ihde, Aaron J., *The Development of Modern Chemistry.* Harper & Row, 1964.

International Cloud Atlas, Vol. 1. Geneva: World Meteorological Organization, 1956.

Jones, Claire, Steve J. Gadler and Paul H. Engstrom, *Pollution: The Air We Breathe.* Lerner Publications Company, 1971.

Kals, W. S., *The Riddle of the Winds.* Doubleday, 1977.

Kay, John, *A Series of Original Portraits and Caricature Etchings.* Adam and Charles Black, 1877.

Keeton, William T., *Biological Science.* W. W. Norton, 1967.

Kellogg, William W., and Robert Schware, *Climate*

Change and Society: Consequences of Increasing Atmospheric Carbon Dioxide. Westview Press, 1981.

Kirk, Ruth, Snow. William Morrow, 1978.

Lamb, H. H., Climate: Present, Past and Future, Vol. 1. London: Methuen & Co., Ltd., 1972.

Lorenz, Edward N., The Nature and Theory of the General Circulation of the Atmosphere. Geneva: World Meteorological Organization, 1967.

Macorini, Edgardo, ed., Strumenti Scientifici del Museo di Storia della Scienza di Firenze. Firenze: Arnoldo Mondadori, 1968.

Middleton, W. E. Knowles:
A History of the Theories of Rain and Other Forms of Precipitation. Franklin Watts, 1966.
Invention of the Meteorological Instruments. Johns Hopkins Press, 1969.

Miller, Albert, and Jack Thompson, Meteorology. Charles E. Merrill, 1979.

Naar, Jon, The New Wind Power. Penguin Books, 1982.

National Research Council:
Energy and Climate. National Academy of Sciences, 1977.
The Upper Atmosphere and Magnetosphere. National Academy of Sciences, 1977.

Navarra, John Gabriel, Atmosphere, Weather and Climate: An Introduction to Meteorology, W. B. Saunders, 1979.

Partington, J. R., A Short History of Chemistry. Harper, 1960.

Pilkington, Roger, The Ways of the Air. Criterion Books, 1962.

Ratcliffe, J. A., Sun, Earth and Radio: An Introduction to the Ionosphere and Magnetosphere. McGraw-Hill, 1970.

Reiter, Elmar R., Jet Streams. Doubleday, 1967.

Reynman, L., Wetterbuchlein. Berlin: A. Asher & Co., 1893.

Riehl, Herbert, Introduction to the Atmosphere. McGraw-Hill, 1978.

Schaefer, Vincent J., and John A. Day, A Field Guide to the Atmosphere. Houghton Mifflin, 1981.

Schneider, Stephen H., The Genesis Strategy: Climate and Global Survival. Plenum Press, 1976.

Scorer, Richard:
Air Pollution. Pergamon Press, 1968.
Clouds of the World: A Complete Color Encyclopedia. Stackpole Books, 1972.
Pollution in the Air: Problems, Policies and Priorities. Routledge & Kegan Paul, 1973.

Shaw, Sir Napier, Manual of Meteorology, Vol. 1, Meteorology in History. Cambridge: Cambridge University Press, 1926.

Skinner, Brian J., ed., Climates Past and Present. William Kaufmann, 1981.

Sootin, Harry, The Long Search: Man Learns about the Nature of Air. W. W. Norton, 1967.

Sproull, Wayne T., Air Pollution and Its Control. Exposition Press, 1972.

Williamson, Samuel J., Fundamentals of Air Pollution. Addison-Wesley Publishing Co., 1973.

Young, Louise B., Earth's Aura. Alfred A. Knopf, 1977.

Periodicals

Akasofu, Syun-Ichi:
"The Aurora." American Scientist, September-October 1981.
"Aurora Borealis: The Amazing Northern Lights." Alaska Geographic, Vol. 6, No. 2, 1979.

Azarin, Beverly, "In the Wake of the Flying Cloud." Science 81, March 1981.

Baum, Rudy M., "Stratospheric Science Undergoing Change." Chemical and Engineering News, September 13, 1982.

Blench, B.J.R., "Luke Howard and His Contribution to Meteorology." Weather, March 1963.

Bonacina, L.C.W., "An Estimation of the Great London Fog of 5-8 December 1952." Weather, November 1953.

Carter, Malcolm N., "Will the Acropolis Someday Be a Copy?" ARTnews, February 1978.

Chameides, William L., and Douglas D. Davis, "Chemistry in the Troposphere." Chemical and Engineering News,

October 4, 1982.

Douglas, C.K.M., and B. A. and K. H. Stewart, "London Fog of December 5-8, 1952." Meteorological Magazine, March 1953.

Fraser, Alistair B., "To See a Dazzling Festival of Light, Just Raise Your Eyes." Smithsonian, January 1981.

Fraser, Alistair B., and William H. Mach, "Mirages." Scientific American, January 1976.

Garmon, L., and C. Simon, "Ozone Depletion: Estimates Halved." Science News, Vol. 121.

Gorham, Eville, "What to Do about Acid Rain." Technology Review, October 1982.

Gosnell, Marianna, "Ozone—the Trick Containing It Where We Need It." Smithsonian, June 1975.

Graves, C. K., "Rain of Troubles." Science 80, July-August 1980.

Gribbin, John, "Monitoring Halocarbons in the Atmosphere." New Scientist, January 18, 1979.

Gwynne, Peter, "Good and Bad News for the U.S. Environment." New Scientist, September 2, 1982.

Haagen-Smit, A. J.:
"Chemistry and Physiology of Los Angeles Smog." Industrial and Engineering Chemistry, June 1952.
"The Control of Air Pollution in Los Angeles." Engineering and Science, December 1954.

Hansen, J., et al., "Climate Impact of Increasing Atmospheric Carbon Dioxide." Science, August 28, 1981.

"Ice Ages Attributed to Orbit Changes." Science News, December 4, 1976.

Iker, Sam, "The Problem of Nitrous Oxide." Mosaic, January/February 1982.

Kidder, Tracy, "A Blemished Sun?" Science 81, July-August 1981.

Kindley, Mark, "For Eye-in-the-Sky Inventors, Kites Can Be Much More than Toys." Smithsonian, June 1982.

Kocivar, Ben, "Tornado Turbine Reaps Power from the Wind." Popular Science, January 1977.

LaBastille, Anne, "Acid Rain: How Great a Menace?" National Geographic, November 1981.

Likens, Gene E., Richard F. Wright, James N. Galloway and Thomas J. Butler, "Acid Rain." Scientific American, October 1979.

Lorenz, Edward N., "The Circulation of the Atmosphere." American Scientist, December 1966.

Lynch, David K., "Atmospheric Halos." Scientific American, April 1978.

Newell, Reginald E., "The Global Circulation of Atmospheric Pollutants." Scientific American, January 1971.

Nussenzveig, H. Moyses, "The Theory of the Rainbow." Scientific American, April 1977.

Oort, Abraham H., "The Energy Cycle of the Earth." Scientific American, September 1970.

Orton, Vrest, "Wilson Alwyn Bentley." Harvard Magazine, November-December 1981.

Parsons, Willard H., "Volcanoes of Guatemala." The Explorer, Vol. 21, No. 4, Winter 1979.

Pearce, Fred, "The Menace of Acid Rain." New Scientist, August 12, 1982.

Peterson, Ivars, "To Catch a Cloud." Science News, August 28, 1982.

Preston, Richard M., "The Life of a Snow Crystal." Country Journal, December 1981.

Revelle, Roger, "Carbon Dioxide and World Climate." Scientific American, August 1982.

Soberman, Robert K., "Noctilucent Clouds." Scientific American, June 1963.

Toon, Owen B., and James B. Pollack, "Atmospheric Aerosols and Climate." American Scientist, May-June 1980.

Van Allen, James A.:
"Interplanetary Particles and Fields." Scientific American, September 1975.
"Radiation Belts around the Earth." Scientific American, March 1959.

West, Susan, "Acid from Heaven." Science News, February 2, 1980.

Whipple, F.J.W.:
"The High Temperature of the Upper Atmosphere as an Explanation of Zones of Audibility." Nature, February

10, 1923.
"The Propagation of Sound to Great Distances." Quarterly Journal of the Royal Meteorological Society, July 1935.

Wilford, John Noble, "Atop Brutal Mt. Washington Scientists Measure the Nation's Worst Weather." The New York Times, January 13, 1981.

Winkler, Erhard M., "Decay of Stone Monuments & Buildings: The Role of Acid Rain." Technology & Conservation, Spring 1982.

Wolkomir, Richard, "Weathermen's Lab Is a Rocky Mountain High." Smithsonian, April 1982.

Ziemann, Hans Heinrich, "Breathless in Athens." GEO, March 1981.

Other Publications

"Acidification Today and Tomorrow." A Swedish study prepared for the 1982 Stockholm Conference on the Acidification of the Environment, Ministry of Agriculture, 1982.

"Acid Precipitation." National Oceanic and Atmospheric Administration, Current Issue Outline 81-1, Washington, D.C., February 1981.

"Acid Rain." Department of Fisheries and Oceans, Ottawa, Ontario, 1980.

Cheng, Robert J.:
"Emissions from Electric Power Plants and Their Impact on the Environment." Atmospheric Sciences Research Center, State University of New York at Albany, no date.
"Physical Properties of Atmospheric Particulates." Atmospheric Sciences Research Center, State University of New York at Albany; also published in Light Scattering by Irregularly Shaped Particles, Donald W. Schuerman, ed., Plenum Publishing, 1980.

Eddy, John, "Climate and the Changing Sun." 1979 Yearbook of Science and the Future, Encyclopedia Britannica, 1978.

Environmental Protection Agency:
"Cleaning the Air: EPA's Program for Air Pollution Control." June 1979.
"Controlling Particulate Emissions for Coal-Fired Boilers." June 1979.
"National Accomplishments in Pollution Control: 1970-1980. Some Case Histories." December 1980.
"Trends in the Quality of the Nation's Air—A Report to the People." October 1980.

"Fact Sheet on Acid Rain." Canadian Embassy, Washington, D.C., no date.

"The Global 2000 Report to the Present: Entering the Twenty-first Century." A report prepared by the Council on Environmental Quality and the Department of State, Washington, D.C., no date.

Kellogg, William W., and Stephen H. Schneider, "Global Air Pollution and Climate Change." IEEE Transactions on Geoscience Electronics, January 1978.

Nonfarmale, Ottorino, "A Method of Consolidation and Restoration of Decayed Sandstones." The Conservation of Stone. Proceedings of the International Symposium, Bologna, June 19-21, 1975. Bologna: Centro Conservazione Sculture all'Aperto.

Panofsky, Hans A., "Threats to the Ozone Layer." Paper presented at the Annual Meeting of the American Association for the Advancement of Science, Washington, D.C., January 6, 1982.

Rowland, F. Sherwood, "Stratospheric Ozone: Earth's Fragile Shield." 1979 Yearbook of Science and the Future, Encyclopedia Britannica, 1978.

Schefter, James L., "Capturing Energy from the Wind." National Aeronautics and Space Administration, NASA SP-455, Washington, D.C., 1982.

Teisserenc de Bort, L., and L. Rotch, "Exposé Technique de l'Étude de l'Atmosphère Marine par Sondages Aériens." Travaux Scientifiques de l'Observatoire de Trappes, Vol. 4, 1909.

Torraca, Giorgio, "Treatment of Stone in Monuments: A Review of Principles and Processes." The Conservation of Stone. Proceedings of the International Symposium, Bologna, June 19-21, 1975. Bologna: Centro Conservazione Sculture all'Aperto.

PICTURE CREDITS

The sources for the illustrations that appear in this book are listed below. Credits from left to right are separated by semicolons, from top to bottom by dashes.

Cover: © Wilhelm Schmidt from Masterfile. 6, 7: Loren A. McIntyre. 8, 9: James A. Sugar from Black Star. 10, 11: © 1977 Lee Snyder/Geophysical Institute/University of Alaska. 12, 13: Dr. Alistair B. Fraser. 14: Dr. Takasi Oguti, Tokyo. 17: The Bettmann Archive. 19: Enrico Genovese, Mondadori, Milan. 20: Franca Principe, Istituto e Museo di Storia della Scienza, Florence, except center, Enrico Genovese, Mondadori, Milan. 21: Franca Principe, Istituto e Museo di Storia della Scienza, Florence, except upper right, Enrico Genovese, Mondadori, Milan (2). 24, 25: Mary Evans Picture Library, London. 26: Library of Congress. 27: By courtesy of The National Portrait Gallery, London—Library of Congress. 28: Library of Congress. 29: Archives Tallandier, Private Collection, Paris. 31: Courtesy Die Chemischen Institute der Universität Heidelberg, photographed by Erwin Böhm, Mainz—art by Kathy Rebeiz—from *Gesammelte Abhandlungen von Robert Bunsen*, Vol. 2, Leipzig, 1904, courtesy Deutsches Museum, Munich, photographed by Erwin Böhm, Mainz. 32: Mary Evans Picture Library, London. 33: Art by Greg Harlin. 34, 35: © 1982 Jim Brandenburg. 36, 37: © 1982 Myrleen Ferguson; Russ Kinne from Photo Researchers—Bill Deane—Paul Drummond from Bryan and Cherry Alexander, Arundel, Sussex, England. 38, 39: Prof. Richard Scorer, London; Shiro Shirahata, Tokyo—© Dr. Alistair B. Fraser (2). 40, 41: Tom Pantages (4)—Specola Vaticana, Citta' del Vaticano, Rome (4). 42: NASA. 47: Brown University Library—Deutsches Museum, Munich. 48, 49: From *Travels in the Air* by James Glaisher, 1871. 51: Su Gooders from Ardea,

London—art by Carol Schwartz. 54: C. G. Robins, courtesy Meteorological Office Library, Bracknell, Berkshire, England. 56: NASA. 57: NASA—art by Greg Harlin. 58, 59: Art by Carol Schwartz. 60, 61: NASA; art by Greg Harlin. 62, 63: Art by Lloyd K. Townsend. 64, 65: Graph by Walter Hilmers; art by Lloyd K. Townsend. 66: *The Miami Herald*. 68: Library of Congress. 70, 71: Jean Dufour, Avignon. 73: Courtesy The Royal Society, London—courtesy The Royal Institution, photographed by Derek Bayes, London. 77: Enertech Corp.; Robert W. Madden. 78, 79: NASA, inset, Ronald Holle. 80: Reinhard Eisele, Augsburg, Federal Republic of Germany. 82: Art by I'Ann Blanchette. 84, 85: Art by Rob Wood. 86: Art by Greg Harlin. 88, 89: © 1982 Thomas Braise, inset, © Georg Gerster from Photo Researchers; inset, NASA. 90, 91: Greg Davis, Tokyo, inset, © Eiji Miyasawa from Black Star. 92: Courtesy Dr. Okitsugu Furuya, Tetra Tech, Inc., photographed by Ken Veeder—Craig Kavafes/Grumman Aerospace. 93: Craig Kavafes/Grumman Aerospace. 94: © 1982 Gary Ladd. 96: Reproduced by kind permission of The Royal Meteorological Society, Bracknell, Berkshire, England, photographed by Derek Bayes. 97: Science Museum, London, reproduced by kind permission of The Royal Meteorological Society, Bracknell, Berkshire, England. 99: Art by Rob Wood. 100: The Mansell Collection, London. 101: Art by Frederic F. Bigio from B-C Graphics. 103: © Bill Brooks from Masterfile; © 1976 Henry Lansford. 104: Reprinted courtesy *The Boston Globe*—NOAA. 106, 107: © Peter Kresan, inset, art by I'Ann Blanchette. 108: Art by I'Ann Blanchette—Ronald Holle. 110: Foto Schapowalow/Ligges, Hamburg. 111: © 1981 Michael Philip Manheim from Photo Researchers. 113: © Peter Kresan. 114: David Falconer. 115-117: © Peter Kresan. 118: Art by Greg Harlin. 120,

121: R. K. Pilsbury from Bruce Coleman, Ltd., Middlesex, England. 122, 123: R. K. Pilsbury from Bruce Coleman Ltd., Middlesex, England—Marvullo from The Image Bank—R. K. Pilsbury from the Natural History Picture Agency, London; Kelly Redmond. 124: Loren A. McIntyre—Dr. James W. A. Fullmer. 125: D. Lepper from The Image Bank. 126, 127: Stephen J. Krasemann/DRK Photo. 128: Francis A. Schiermeier. 131: Roger J. Cheng, State University of New York, Albany. 132: William Buell. 134: BBC Hulton Picture Library, London. 137: Art by I'Ann Blanchette; George C. Atamian/Talcott Mountain Science Center—Catherine Azema from Pitch, Paris. 138: © Freeman Patterson from Masterfile. 139: Art by Frederic F. Bigio from B-C Graphics. 141: © Jerry Young, London. 142, 143: Art by Lloyd K. Townsend. 145: © Ted Spiegel from Black Star. 146: Art by I'Ann Blanchette. 147: © 1980 Roger J. Cheng, State University of New York, Albany. 148, 149: Wilfried Bauer, Hamburg. 150, 151: Wilfried Bauer, Hamburg; Giorgio Lotti, Milan (2). 152, 153: Giorgio Lotti, Milan—Ernesta Vergani, Treviso, Italy (2); Antonio Guerra, Villani e Figli, Bologna. 154: Wayne Scorce/Visions. 156: Peter Kresan, courtesy The University of Arizona Laboratory of Tree-Ring Research. 157: Deep Sea Drilling Project/Scripps Institution of Oceanography. 158: Museum of London. 159: Photo Flammarion, Paris; Photo Madeleine Leroy-Ladurie, Paris. 162: Paul Chesley/ASPEN. 163: The National Center for Atmospheric Research, sponsored by The National Science Foundation, except bottom left, Paul Chesley/ASPEN. 164: Art by Kathy Rebeiz. 165: Graph by Walter Hilmers. 166, 167: NASA. 168: American Science and Engineering Inc., courtesy Harvard College Observatory. 170, 171: Daniele Pellegrini, Milan.

INDEX

Time-Life Books Inc. offers a wide range of fine recordings, including a *Country & Western Classics* series. For subscription information, call 1-800-621-8200, or write TIME-LIFE RECORDS, Time & Life Building, Chicago, Illinois 60611.